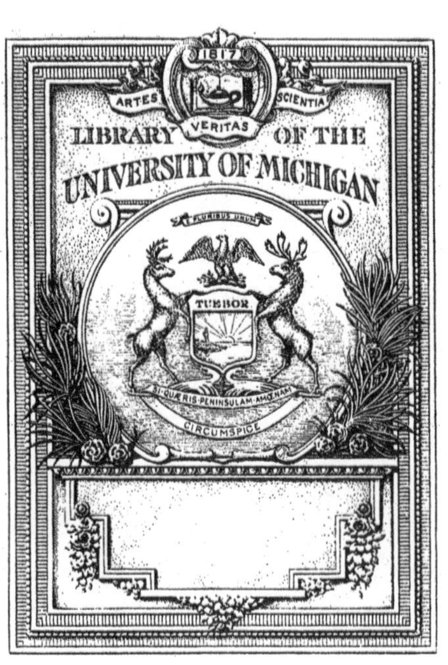
1817
ARTES
SCIENTIA
VERITAS
LIBRARY OF THE
UNIVERSITY OF MICHIGAN
TUEBOR
SI·QUÆRIS·PENINSULAM·AMŒNAM
CIRCUMSPICE

AN ELEMENTARY TREATISE

ON

CUBIC AND QUARTIC CURVES

AN ELEMENTARY TREATISE

ON

CUBIC AND QUARTIC CURVES

BY

A. B. BASSET, M.A., F.R.S.

TRINITY COLLEGE CAMBRIDGE

CAMBRIDGE

DEIGHTON BELL AND CO.

LONDON GEORGE BELL AND SONS

1901

Cambridge

PRINTED BY J. AND C. F. CLAY

AT THE UNIVERSITY PRESS

PREFACE.

THE present work originated in certain notes, made about twenty-five years ago, upon the properties of some of the best-known higher plane curves; but upon attempting to revise them for the press, it appeared to me impossible to discuss the subject adequately without investigating the theory of the singularities of algebraic curves. I have accordingly included Plücker's equations, which determine the number and the species of the simple singularities of any algebraic curve; and have also considered all the compound singularities which a quartic curve can possess.

This treatise is intended to be an elementary one on the subject. I have therefore avoided investigations which would require a knowledge of Modern Algebra, such as the theory of the invariants, covariants and other concomitants of a ternary quantic; and have assumed scarcely any further knowledge of analysis on the part of the reader, than is to be found in most of the ordinary text-books on the Differential Calculus and on Analytical Geometry. I have also endeavoured to give special prominence to geometrical methods, since the experience of many years has convinced me that a judicious combination of geometry and analysis is frequently capable of shortening and simplifying, what would otherwise be a tedious and lengthy investigation.

The introductory Chapter contains a few algebraic definitions and propositions which are required in subsequent portions of the work. The second one deals with the elementary theory of the singularities of algebraic curves and the theory of polar curves. The third Chapter commences with an explanation of tangential coordinates and their uses, and then proceeds to discuss a variety of miscellaneous propositions connected with reciprocal polars, the circular points at infinity and the foci of curves. Chapter IV is devoted to Plücker's equations; whilst Chapter V contains an account of the general theory of cubic curves, including the formal proof of the principal properties which are common to all curves of this degree. In this Chapter I have almost exclusively employed trilinear coordinates, since the introduction of a triangle of reference, whose elements can be chosen at pleasure, constitutes a vast improvement on the antiquated methods of homogeneous coordinates and abridged notation. Chapter VI is devoted to the consideration of a variety of special cubics, including the particular class of circular cubics which are the inverses of conics with respect to their vertices; and in this Chapter the method of Cartesian coordinates is the most appropriate. A short Chapter then follows on curves of the third class, after which the discussion of quartic curves commences.

To adequately consider such an extensive subject as quartic curves would require a separate treatise. I have therefore confined the discussion to the simple and compound singularities of curves of this degree, together with a few miscellaneous propositions; and in Chapter IX, I proceed to investigate the theory of bicircular quartics and cartesians, concluding with the general theory of circular cubics, which is better treated as a particular case of bicircular quartics than as a special case of cubic curves. Chapter X is devoted to the consideration of various well known quartic curves, most of which are bicircular or are cartesians; whilst Chapter XI deals with cycloidal curves, together with a few miscellaneous curves which frequently occur in mathematical investigations. The theory of projection, which forms the subject of the last Chapter, is explained in most treatises on Conics; but except in the case of conics, due weight has not always been given to the important fact that the projective properties of any special class of curves can be deduced from those of the simplest curve of the species. Thus all the projective properties of tricuspidal quartics can be obtained from those of the three-cusped hypocycloid or the cardioid; those of quartics with a node and a pair of cusps from the limaçon; those of quartics with three biflecnodes from the lemniscate of Bernoulli or the reciprocal polar of the four-cusped hypocycloid; whilst the properties of binodal and bi-cuspidal quartics can be obtained from those of bicircular quartics and cartesians.

Whenever the medical profession require a new word they usually have recourse to the Greek language, and mathematicians would do well to follow their example; since the choice of a suitable Greek word supplies a concise and pointed mode of expression which saves a great deal of circumlocution and verbosity. The old-fashioned phrase "a non-singular cubic or quartic curve" involves a contradiction of terms, since Plücker has shown that all algebraic curves except conics possess singularities; and I have therefore introduced the words *autotomic* and *anautotomic* to designate curves which respectively do and do not possess multiple points. The words *perigraphic, endodromic* and *exodromic*, which are defined on page 14, are also useful; in fact a word such as *aperigraphic* is indispensable in order to avoid the verbose phrase "a curve which has branches extending to infinity."

At the present day the subject of Analytical Geometry covers so vast a field that it is by no means easy to decide what to insert and what to leave out. I trust, however, that the present work will form a useful introduction to the higher branches of the subject; and will facilitate the study of a variety of curves whose properties, by reason of their beauty and elegance, deserve at least as much attention as the well-worn properties of conics.

FLEDBOROUGH HALL,
HOLYPORT, BERKS.
August, 1901.

CONTENTS.

CHAPTER I.

INTRODUCTION.

CHAPTER II.

THEORY OF CURVES.

CHAPTER III.

TANGENTIAL COORDINATES.

CHAPTER IV.

PLÜCKER'S EQUATIONS.

CHAPTER V.

CUBIC CURVES.

CHAPTER VI.

SPECIAL CUBICS.

CHAPTER VII.

CURVES OF THE THIRD CLASS.

CHAPTER VIII.

QUARTIC CURVES.

CHAPTER IX.

BICIRCULAR QUARTICS.

CHAPTER X.

SPECIAL QUARTICS.

CHAPTER XI.

MISCELLANEOUS CURVES.

CHAPTER XII.

THEORY OF PROJECTION.

ADDENDA AND CORRIGENDA.

CHAPTER ·I.

INTRODUCTION.

1. BEFORE commencing the study of plane curves, we shall prove certain propositions connected with the Algebra of Quantics which will be required in subsequent Chapters of this work.

A homogeneous function of any number of variables is called a *quantic*. Quantics are called *binary, ternary, quaternary, n*-ary according as they contain two, three, four or n variables ; whilst the degree of the quantic is denoted by the words *quadric, cubic, quartic, n*-tic. Thus the general equation in Cartesian coordinates of three straight lines through the origin is a *binary cubic*, whilst the general equation of a conic in trilinear coordinates is a *ternary quadric*.

The most general expression for a binary quantic is

$$a_0 x^n + a_1 x^{n-1} y + a_2 x^{n-2} y^2 + \dots a_{n-1} x y^{n-1} + a_n y^n \dots\dots(1),$$

which is usually written in the abridged form $(a_0, a_1, \dots a_n \!\!\ \rangle\!\langle\ x, y)^n$. This form is generally the most convenient to employ in geometrical investigations ; but in purely analytical ones it is usually better to use the form

$$a_0 x^n + n a_1 x^{n-1} y + \frac{n(n-1)}{2!} a_2 x^{n-2} y^2 + \dots n a_{n-1} x y^{n-1} + a_n y^n \dots(2),$$

in which each term is multiplied by the coefficients of x in the expansion of $(1+x)^n$. In this expression the coefficients are said to be *binomial*, and the quantic is denoted by

$$(a_0, a_1, a_2, \dots a_n \!\!\ \rangle\!\langle\ x, y)^n.$$

For quantics containing more than two variables a similar notation is employed. Thus the expression

$$(a_0, a_1, \dots a_q \!\!\ \rangle\!\langle\ x_1, x_2, \dots x_p)^n$$

denotes a p-ary n-tic, in which the different terms are multiplied by the coefficients of the corresponding term in the expansion of

$$(x_1 + x_2 + \ldots x_p)^n.$$

2. If F be a quantic, the result of eliminating the variables $x_1, x_2, \ldots x_n$ between the equations

$$\frac{dF}{dx_1} = 0, \quad \frac{dF}{dx_2} = 0, \ldots \frac{dF}{dx_n} = 0$$

is called the *discriminant* of the quantic.

The discriminant of every quadric can be at once written down in the form of a symmetrical determinant; for in this case dF/dx_1 &c. are linear functions of the variables. Thus the discriminant of the ternary quadric

$$ax^2 + by^2 + cz^2 + 2fyz + 2gzx + 2hxy \quad \ldots\ldots\ldots\ldots(3)$$

is
$$\begin{vmatrix} a, & h, & g \\ h, & b, & f \\ g, & f, & c \end{vmatrix}$$

or
$$abc + 2fgh - af^2 - bg^2 - ch^2 \ldots\ldots\ldots\ldots\ldots(4),$$

which expresses the condition that the quadric should be resolvable into two linear factors. When the quantic is not a quadric, the elimination must be performed by the methods explained in treatises on Algebra; but for binary cubics and quartics, the elimination may easily be performed by the following process.

3. Let

$$(a_0, a_1, \ldots a_n \unrhd x, y)^n = 0 \text{ and } (b_0, b_1, \ldots b_n \unrhd x, y)^n = 0$$

be two binary n-tics. If both these equations be divided by y^n and $z = x/y$, they become two equations of the nth degree in z. Multiply the first equation by b_n, and the second by a_n, subtract and divide out by z, and the resulting equation is one of degree $n - 1$. Multiply the first equation by b_0 and the second by a_0, and subtract, and the resulting equation will also be of degree $n - 1$. We have therefore replaced the two equations of degree n by two other equations of degree $n - 1$, and the process may be continued until we arrive at two simple equations from which z can be eliminated.

The result of eliminating x and y between two binary quantics is called their *eliminant*. Eliminants are sometimes called resultants; but the former term is the better one, since it is more expressive of the precise nature of the process employed.

4. We shall now write down the discriminants Δ of the binary quadric, cubic and quartic.

(i) The quadric $\quad (a, b, c \,\, x, y)^2$
$$\Delta = ac - b^2 \quad\dots\dots\dots\dots\dots\dots(5).$$

(ii) The cubic $\quad (a, b, c, d \,\, x, y)^3$
$$\Delta = (ad - bc)^2 - 4(ac - b^2)(bd - c^2) \quad\dots\dots\dots(6),$$
or $\qquad\quad \Delta = a^2 d^2 - 6abcd + 4ac^3 + 4b^3 d - 3b^2 c^2 \dots\dots\dots(7).$

(iii) The quartic $\quad (a, b, c, d, e \,\, x, y)^4$
$$\Delta = I^3 - 27 J^2 \quad\dots\dots\dots\dots\dots(8),$$
where
$$\left.\begin{aligned} I &= ae - 4bd + 3c^2 \\ J &= ace + 2bcd - ad^2 - b^2 e - c^3 \end{aligned}\right\} \quad\dots\dots\dots\dots(9).$$

5. We shall next establish certain propositions concerning the roots of an equation. These theorems are contained in most treatises on the Theory of Equations, but it will be convenient to collect them for future reference.

The condition that the equation $F(z) = 0$ should have r equal roots is obtained by eliminating z between the r equations.

$$F(z) = 0, \;\; F'(z) = 0, \;\; \dots \;\; F^{r-1}(z) = 0.$$

Let α be one of the roots of $F(z) = 0$; and let $z - \alpha = h$; then by Taylor's theorem

$$F(z) = F(\alpha + h) = F(\alpha) + hF'(\alpha) + \tfrac{1}{2}h^2 F''(\alpha) + \dots = 0.$$

Since α is a root of the equation, $F(\alpha) = 0$; whence dividing out by h, it follows that if a second root is equal to α, $F'(\alpha) = 0$. Continuing this process it follows that if r roots are equal to α, all the differential coefficients of $F(\alpha)$ up to the $(r-1)$th must vanish.

6. *The condition that the equation $(a_0, a_1, \dots a_n \,\, z, 1)^n = 0$ should have two equal roots is that the discriminant of the binary quantic $(a_0, a_1, \dots a_n \,\, x, y)^n$ should vanish.*

Let
$$F(z) = (a_0, a_1, \dots a_n \,\, z, 1)^n \dots\dots\dots\dots(10),$$
$$F(x, y) = (a_0, a_1, \dots a_n \,\, x, y)^n \dots\dots\dots\dots(11),$$

then we have already shown that the condition that (10) should have a pair of equal roots is that the eliminant of $F(z) = 0$ and $F'(z) = 0$ should vanish; and consequently the eliminant of $F'(z) = 0$ and $nF(z) - zF'(z) = 0$ must vanish. But on writing out these expressions in full, it will be found that these conditions are the same as

$$\frac{dF}{dx} = 0, \quad \frac{dF}{dy} = 0,$$

which are the conditions that the discriminant of $F(x, y)$ should vanish.

If $F(x, y)$ is a binary n-tic given by (11), it follows that $d^r F / dx^r$ divided by the numerical factor $n(n-1)\ldots(n-r+1)$ is a binary $(n-r)$-tic; in other words if

$$F(x, y) = (a_0, a_1, \ldots a_n \rangle x, y)^n,$$

then
$$\frac{1}{n(n-1)\ldots(n-r+1)} \frac{d^r F}{dx^r} = (a_0, a_1, \ldots a_{n-r} \rangle x, y)^{n-r},$$

whence the theorem of § 5 may be otherwise stated :—*If*

$$F(z) = (a_0, a_1, \ldots a_n \rangle z, 1)^n,$$

the condition that the equation $F(z) = 0$ may have r equal roots is that the discriminants of $F(z)$ and all its differential coefficients up to the $(r-2)$th should vanish.

7. We shall proceed to find the conditions for equalities between the roots of cubic and quartic equations.

The Cubic.

The condition that two of the roots of the cubic

$$(a, b, c, d \rangle z, 1)^3 = 0$$

should be equal is that the discriminant should vanish; whence by (6) the required condition is

$$(ad - bc)^2 - 4(ac - b^2)(bd - c^2) = 0 \quad \ldots\ldots\ldots(12).$$

If a third root is equal the discriminant of the quadratic $(a, b, c \rangle z, 1)^2 = 0$ must also vanish; whence by (5) we have the second condition

$$ac - b^2 = 0.$$

Combining this with (12) we obtain

$$a/b = b/c = c/d \quad \dots\dots\dots\dots\dots(13),$$

which are the required conditions that three roots should be equal.

The Quartic.

The conditions that three of the roots of the quartic

$$(a,\ b,\ c,\ d,\ e\ \text{⌇}\ z,\ 1)^4 = 0$$

should be equal is that the discriminants (6) and (8) should vanish. From (6) and (9) we obtain

$$I = ae - c^2 - 4\,(bd - c^2)$$
$$= ae - c^2 - \frac{(ad - bc)^2}{ac - b^2}$$
$$= \frac{aJ}{ac - b^2},$$

which by (8) requires that

$$I = 0, \quad J = 0 \ \dots\dots\dots\dots\dots(14).$$

The condition that the quartic should have four equal roots involves the additional equation $ac - b^2 = 0$. In combination with (14), this leads to the three equations

$$a/b = b/c = c/d = d/e \quad \dots\dots\dots\dots(15).$$

The conditions that a quartic should have two pairs of equal roots, which are the conditions that the quartic should be the square of a quadratic factor, can be readily obtained by means of the relations which exist between the coefficients and the two pairs of equal roots α and β. These four relations, after α and β have been eliminated, lead to the results

$$\left.\begin{array}{c} ad^2 = b^2 e \\ 2b^3 + a^2 d = 3abc \end{array}\right\} \quad \dots\dots\dots\dots(16).$$

8. The condition that the product of two of the roots of the equation $F(z) = 0$ should be equal to -1 is the condition that the eliminant of the equations $F(z) = 0$ and $F(-z^{-1}) = 0$ should vanish. This eliminant is required in finding the orthoptic locus of a curve, and we shall show how it can be obtained in the case

of the quartic $(a, b, c, d, e \text{\c{}} z, 1)^4 = 0$. Express the coefficients in terms of the four roots α, α^{-1}, β, γ, and it will be found on eliminating α that the four relations can be reduced to three which are functions of $\beta + \gamma$ and $\beta\gamma$. Eliminate these two quantities, and the result is

$$(b + d)(ad + be) + (a + c + e)(a - e)^2 = 0 \quad\ldots\ldots\ldots(17).$$

Putting successively $c = 0$, $d = 0$ we obtain the corresponding eliminants for a cubic and a quadratic, which are

$$(b + d) d + (a + c) a = 0 \quad\ldots\ldots\ldots\ldots\ldots(18),$$
$$a + c \quad = 0 \quad\ldots\ldots\ldots\ldots\ldots(19).$$

When $F(z)$ is of the nth degree, the eliminant is of degree $n - 1$ in the coefficients.

The Hessian.

9. Let u be any quantic; let u_1, u_2, u_3 ... denote its first differential coefficients with respect to the variables x_1, x_2, x_3 ...; also let u_{11}, u_{12}, u_{13} ... denote the differential coefficients of u_1 with respect to x_1, x_2, x_3 .... Then the determinant

$$\begin{vmatrix} u_{11}, & u_{12}, & u_{13} \ldots \\ u_{12}, & u_{22}, & u_{23} \ldots \\ u_{13}, & u_{23}, & u_{33} \ldots \\ \ldots\ldots\ldots \\ \ldots\ldots\ldots \end{vmatrix} \quad\ldots\ldots\ldots\ldots\ldots(20)$$

is called the *Hessian* of u. We shall denote it by the letter H.

The Hessian is so called because it was first studied by the German mathematician Hesse, and it has important applications in the theory of curves. In the next Chapter we shall show that any ternary quantic equated to zero is the equation of a curve in trilinear coordinates; and that the Hessian when equated to zero represents a curve which passes through the points of inflexion of the original curve.

The Hessian of a ternary quantic is evidently obtained by substituting the values of $u_{11} \ldots$, $u_{23} \ldots$ for $a \ldots$, $f \ldots$ in (4) and is therefore equal to

$$u_{11} u_{22} u_{33} + 2 u_{23} u_{13} u_{12} - u_{11} u^2_{23} - u_{22} u^2_{13} - u_{33} u^2_{12} \ldots\ldots(21),$$

and since u_{11} &c. are of degree $n-2$ in the variables, it follows that the Hessian is of degree $3(n-2)$. Hence the Hessian of a ternary cubic is also a cubic.

The reader who possesses an elementary knowledge of invariants and covariants will observe that discriminants are invariants and Hessians are covariants. Binary quadrics and cubics have only one invariant, viz. their discriminant. Binary quartics have two invariants, viz. the functions I and J, the vanishing of which expresses the condition that the quartic (regarded as an equation in y/x) has three equal roots. A binary cubic has two covariants, viz. its Hessian H, and its cubicovariant G. The values of H and G are

$$H = (ac - b^2)\,x^2 + (ad - bc)\,xy + (bd - c^2)\,y^2 \ldots\ldots\ldots\ldots(22),$$

$$G = (a^2d - 3abc + 2b^3, \quad abd - 2ac^2 + b^2c,$$

$$-\,acd + 2b^2d - bc^2, \quad -ad^2 + 3bcd - 2c^3 \!\!\gtrdot\!\! x,\ y)^3 \ldots(23).$$

A ternary cubic has two invariants of degrees 4 and 6 in the coefficients, which are usually denoted by S and T. For further information on this subject the reader may consult Elliott's *Algebra of Quantics*; Salmon's *Lessons in Higher Algebra*; and Salmon's *Higher Plane Curves*.

CHAPTER II.

THEORY OF CURVES.

10. THE general equation of a curve of the nth degree, when expressed in Cartesian coordinates, may be written in the form

$$u_n + u_{n-1} + \ldots\ldots u_1 + u_0 = 0 \ldots\ldots\ldots\ldots\ldots(1),$$

where u_n is a binary quantic in x and y. The number of terms in u_r is obviously one more than in u_{r-1}, and is therefore equal to $r + 1$; whence the total number of terms in (1) is $1+2+\ldots\ldots n+1$, which is equal to $\frac{1}{2}(n+1)(n+2)$.

The number of independent constants in (1) is equal to one less than the number of terms it contains, since the generality of (1) remains unaltered when each term is divided by u_0 and new constants are substituted for the ratios of the old ones to u_0. Hence the general equation of a curve of the nth degree contains only $\frac{1}{2}n(n+3)$ independent constants, and therefore the curve can only be made to satisfy the same number of independent conditions.

The general equation of a curve of the nth degree in trilinear coordinates (α, β, γ) may be written in the form

$$\alpha^n u_0 + \alpha^{n-1}u_1 + \ldots \alpha u_{n-1} + u_n = 0 \ldots\ldots\ldots\ldots(2),$$

where u_n is a binary quantic in β and γ. Hence every ternary quantic of degree n contains $\frac{1}{2}(n+1)(n+2)$ terms, and $\frac{1}{2}n(n+3)$ independent constants.

11. It is shown in treatises on Algebra that if U_m, V_n be any rational algebraic functions of x and y of degrees m and n respectively, and if y be eliminated between the equations $U_m = 0$, $V_n = 0$ the resulting equation in x will be of degree mn. Hence two curves of the mth and nth degrees intersect in mn points. Accordingly a straight line intersects a curve of the nth degree in

n points, and if n is odd one at least of these points must be real. A conic intersects the curve in $2n$ points, of which all must be imaginary or an even number must be real.

Multiple Points.

12. When a curve cuts itself once at the same point, the latter is called a *double* point, and the curve has two tangents at this point. When the two tangents are distinct, the double point is called a *crunode* or shortly a *node*; when they are imaginary, the point is called an *acnode* or a *conjugate point*; and when they are coincident, the point is called a *spinode* or *cusp*.

An example of the three kinds of double points is furnished by the limaçon, whose equations in polar and Cartesian coordinates are

$$r = a + b \cos \theta$$

and
$$(x^2 + y^2)^2 - 2bx \, (x^2 + y^2) = (a^2 - b^2) \, x^2 + a^2 y^2 \ \ldots\ldots(3).$$

When $b > a$, the curve is the inverse of a hyperbola with respect to a focus, and cuts itself at the origin. The angle between the tangents to the two branches is equal to $2 \cos^{-1} a/b$, and the origin is a *node*. When $a = b$, the loop disappears, and the origin becomes a *cusp*, in which case the curve is called a *cardioid*. When $a > b$, the Cartesian equation is satisfied by $x = 0$, $y = 0$, but no real branches of the curve pass through the origin. The tangents at the origin are therefore imaginary, and the latter is a *conjugate point*. It thus appears that conjugate points are isolated points whose coordinates satisfy the equation of the curve, but the curve itself does not pass through them.

When a curve possesses a conjugate point, the latter is always the limit of an oval which shrinks up into a point; and it will be shown hereafter that the acnodal limaçon is a limiting form of a quartic curve, called the oval of Descartes, which consists of two ovals, one of which lies inside the other.

13. When three branches of a curve pass through a point, the latter is called a *triple* point; and generally if k distinct branches of a curve pass through a point, the latter is called a *multiple point of order k*.

Contact.

14. When two curves intersect one another in $r + 1$ coincident points, they are said to have a *contact of the rth order* with one another.

When two curves have a contact of the first order with one another at two distinct points, they are said to have a *double contact* with one another.

15. *Every straight line through a double point has a contact of the first order with the curve at that point.*

Let *any* two points be taken on the curve in the neighbourhood of a double point; then the straight line through these points intersects the curve in at least two points. Accordingly by making the two points move up to coincidence with the double point, it follows that every straight line through the latter intersects the curve in two coincident points.

16. *Every tangent at a double point has a contact of the second order with the curve.*

Take any point P on the curve near the double point; then the line through P and the double point ultimately becomes a tangent at the latter. But since every line through a double point intersects the curve in two coincident points, the tangent at the double point intersects the curve in three coincident points.

In the same way it can be shown that if a curve A passes through a double point on a curve B, the former has a contact of the first order with the latter at the double point, but the curves will not *touch* one another unless the curve A intersects one of the branches of B, which passes through the double point, in two coincident points. In this case the curve A will have a contact of the first order *with the particular branch*, and a contact of the second order *with the curve B* at the double point.

17. A tangent to a curve is usually defined as a line which intersects the curve in two coincident points, or as a line which has a contact of the first order with the curve; but this definition is only applicable to curves of the second degree. For we have just shown that every line through a double point satisfies the preceding definition of tangency, whereas there are only *two*

tangents at a double point, both of which have a contact of the second order with the curve. The preceding definition is consequently wanting in accuracy; and we shall therefore define a tangent at any point of a curve as *the line of closest possible contact with the curve at that point.*

We shall now resume the consideration of multiple points.

18. *If a curve be referred to a point on itself as origin, the linear term equated to zero is the equation of the tangent at the origin.*

The general equation of a curve of the nth degree when expressed in polar coordinates may be written

$$A + (B\cos\theta + C\sin\theta)\,r + (D\cos^2\theta + E\sin 2\theta + F\sin^2\theta)\,r^2$$
$$+\, u_3 + \ldots u_n = 0 \ldots\ldots\ldots\ldots\ldots\ldots(4).$$

When the origin lies on the curve $A = 0$, and one value of r is zero; if, however, θ be determined so that

$$B\cos\theta + C\sin\theta = 0 \ldots\ldots\ldots\ldots\ldots (5),$$

two values of r will be zero, and the line

$$Bx + Cy = 0$$

is the tangent to the curve at the origin.

19. *If the origin be a double point the term of lowest dimensions is the quadratic term, and this term equated to zero is the equation of the tangents at the double point.*

If $B = C = 0$ two values of r will be zero whatever the value of θ may be; and every line passing through the origin will have a contact of the first order with the curve. If, however, θ be determined so that

$$D\cos^2\theta + E\sin 2\theta + F\sin^2\theta = 0 \ldots\ldots\ldots\ldots(6),$$

three values of r will be zero, and the two lines whose inclinations to the axis of x are determined by (6) will have a contact of the second order with the curve. The origin is therefore a double point, and (6) gives the directions of the tangents at the origin. Their equation is

$$Dx^2 + 2Exy + Fy^2 = 0 \ldots\ldots\ldots\ldots\ldots(7).$$

It appears from (7) that the two tangents at the double point will be real, coincident or imaginary according as $E^2 >$ or $=$ or $< DF$, in which three respective cases the origin will be a node, a cusp or a conjugate point. In the case of the limaçon, the tangents at the origin are given by the equation

$$(a^2 - b^2)\, x^2 + a^2 y^2 = 0,$$

hence the latter is a node or a conjugate point according as $b >$ or $< a$, that is according as the limaçon is hyperbolic or elliptic. When $a = b$, the curve is a cardioid and the origin is a cusp whose cuspidal tangent is the axis of x.

Although the cusp has occurred as a species of double point, it may be well to remark that it is really a distinct singularity; moreover there are different kinds of cusps, such as *rhamphoid*, *tacnode* and *oscnode* cusps, all of which excepting the spinode are multiple points of a higher order than the second. The spinode is sometimes called a keratoid cusp from a fancied resemblance to the form of a horn.

If $D = E = F = 0$, the origin is a triple point, the tangents at which are determined by the equation $u_3 = 0$. There are four kinds of triple points according as the roots of this equation are (i) real and unequal, (ii) real and two equal, (iii) real and all three equal, (iv) one real and two complex.

Since every line straight or curved which passes through a double point intersects the curve in two coincident points, it follows that a cubic cannot have more than one double point; for if it had two, the line joining them would intersect the cubic in four points, which is impossible. Similarly a quartic cannot have more than three double points; for if it had four, a conic could be described through the four double points and any fifth point on the curve, and the conic would therefore intersect the quartic in nine points, which is impossible, since a conic and a quartic cannot intersect in more than eight points. Accordingly a limit exists to the number of double points which a curve can have. A curve may also have a lower number of double points than the maximum; and it will be shown hereafter that unless the discriminant of a ternary quantic vanishes, the curve obtained by equating the quantic to zero has no double points. A curve may also have imaginary multiple points, which must be reckoned amongst the singularities in the same way as real multiple points. For example, we shall prove later on that the cardioid has three cusps, one of which is real and the other two imaginary.

The foregoing remarks only apply to equations which represent *proper* curves, that is to say equations which are incapable of being resolved into factors which represent two or more curves of a lower degree than that of the equation. For example, if a cubic equation be capable of resolution into a linear and a quadric factor, the two points of intersection of the straight line and the conic satisfy the analytical conditions of a double point; and by parity of reasoning it appears that whenever a curve of the nth degree has more than the maximum number of double points, the equation representing the curve breaks up into factors, each of which represents a curve of a lower degree than the nth.

20. Before proceeding further it will be desirable to give a few definitions.

The *deficiency D* of a curve is the number by which the number of double points, real or imaginary, falls short of the maximum.

The *class* of a curve is the number of tangents, real or imaginary, which can be drawn from any point to the curve. We shall denote the class by the letter m.

A *point of inflexion* is a point, which is not a double point, where the tangent has a contact of the second order with the curve. The tangent at a point of inflexion is sometimes called a *stationary tangent*.

A *point of undulation* is a point, which is not a triple point, where the tangent has a contact of the third order with the curve. No curve of a lower degree than a quartic can have points of undulation.

A *double tangent* is a line which touches a curve at two distinct points. Since a double tangent intersects a curve in four points, no curve of a lower degree than a quartic can have a double tangent; but curves of a higher degree than the fourth may have multiple tangents of a higher order. Also a multiple tangent may have a contact of a higher order than the first. Thus a sextic may have (i) a triple tangent having a contact of the first order at three distinct points, (ii) a double tangent touching the curve at a point of undulation and at a point at which the contact is of the first order, (iii) a double tangent touching the curve at two points of inflexion.

21. Curves which possess double points will be called *autotomic* (self-cutting); and curves which do not possess these singularities will be called *anautotomic*.

A continuous closed curve will be called a *perigraphic* curve; whilst a curve which possesses branches extending to infinity will be called an *aperigraphic* curve. A circle or an ellipse is the simplest example of a perigraphic curve; whilst parabolas and hyperbolas are aperigraphic. Curves of an even degree may be perigraphic or aperigraphic; but all curves of an odd degree are aperigraphic.

All curves of an even degree, except conics, may consist of two or more perigraphic portions which may lie entirely within or entirely without one another. In the former case the curves will be called *endodromic*, and in the latter *exodromic*.

It will be shown in Chapter X. that the oval of Descartes is an endodromic curve, consisting of two ovals, one of which lies inside the other; whilst, for certain values of the constants, the oval of Cassini is an exodromic curve which consists of two detached ovals external to one another.

A curve which consists of one, two, three, &c. distinct portions, which may or may not be perigraphic, is called *unipartite, bipartite, tripartite*, &c. Thus an oval of Descartes, which has three real collinear foci, is endodromic and bipartite; but a Cartesian, which has one real and two imaginary collinear foci, is unipartite and perigraphic.

Conditions for a Double Point.

22. The equation of any line in trilinear coordinates which passes through the point (f, g, h) may be written in the form

$$(\alpha - f)/l = (\beta - g)/m = (\gamma - h)/n = r \ldots\ldots\ldots\ldots(8),$$

whence $\qquad \alpha = f + lr, \quad \beta = g + mr, \quad \gamma = h + nr \ \ldots\ldots\ldots\ldots(9).$

To find where (8) intersects the curve $F(\alpha, \beta, \gamma) = 0$, substitute the values of α, β, γ from (9) and expand by Taylor's theorem, and we obtain

$$0 = F(f, g, h) + r\left(l\frac{d}{df} + m\frac{d}{dg} + n\frac{d}{dh}\right)F + \tfrac{1}{2}r^2(\ldots) + \&c\ldots(10).$$

If (f, g, h) lies on the curve, $F(f, g, h) = 0$, and one of the values of r will be zero; if however (8) has a contact of the first order with the curve, two of the values of r must be zero, the condition for which is that

$$l\frac{dF}{df} + m\frac{dF}{dg} + n\frac{dF}{dh} = 0 \quad\dots\dots\dots\dots\dots(11).$$

When (8) is a tangent to the curve, the values of (l, m, n) must satisfy (11), which is the condition that (8) should touch the curve; whence substituting the values of (l, m, n) from (8) and taking account of Euler's theorem, we obtain

$$\alpha\frac{dF}{df} + \beta\frac{dF}{dg} + \gamma\frac{dF}{dh} = 0 \quad\dots\dots\dots\dots\dots(12),$$

which is the equation of the tangent at (f, g, h).

If however the point (f, g, h) is such that

$$dF/df = 0, \ dF/dg = 0, \ dF/dh = 0 \quad\dots\dots\dots\dots(13),$$

every line through this point has a contact of the first order with the curve, and the point will be a double point. In this case it will be possible to eliminate (f, g, h) from (13); whence, *the condition that a curve should have a double point is that the discriminant of its equation should vanish.*

Since a pair of intersecting straight lines is the only conic which possesses a double point, it follows that the vanishing of the discriminant is the condition that a ternary quadric should break up into two linear factors.

If a curve be given by the Cartesian equation $F(x, y) = 0$, it can be shown in the same manner that the double points, supposing any exist, are determined by the equations

$$dF/dx = 0, \ dF/dy = 0 \quad\dots\dots\dots\dots\dots(14).$$

It can also be shown that if the equation of the curve be written in the form $\sum_{0}^{n} u_r = 0$, where u_r is a binary quantic in x and y, the condition that double points should exist is that the discriminant of the ternary quantic $\sum_{0}^{n} u_r z^{n-r}$ should vanish.

Polar Curves.

23. Before considering the theory of polar curves, it will be convenient to explain the notation that will be employed, and also to prove a preliminary proposition.

The letters α, β, γ will be employed to denote the trilinear coordinates of a *variable* point; the letters (f, g, h) will denote the coordinates of a fixed point in the plane of the curve; and the letters ξ, η, ζ the coordinates of a fixed point on a curve. Also the letters Δ, Δ' will be used to denote the operators

$$\Delta = f\frac{d}{d\alpha} + g\frac{d}{d\beta} + h\frac{d}{d\gamma},$$

$$\Delta' = \alpha\frac{d}{df} + \beta\frac{d}{dg} + \gamma\frac{d}{dh}.$$

If $F(x, y, z)$ be any ternary quantic of degree n, and if $\alpha+f$, $\beta+g$, $\gamma+h$ be written for x, y, z, then

$$\frac{\Delta^p}{p!}F(\alpha, \beta, \gamma) = \frac{\Delta'^{n-p}}{(n-p)!}F(f, g, h).$$

By Taylor's theorem,

$$F(\alpha+f, \beta+g, \gamma+h) = F(\alpha, \beta, \gamma) + \Delta F + \frac{\Delta^2}{2!}F + \dots \frac{\Delta^n}{n!}F \dots(15),$$

and

$$F(\alpha+f, \beta+g, \gamma+h) = F(f, g, h) + \Delta'F' + \frac{\Delta'^2}{2!}F' + \dots \frac{\Delta'^n}{n!}F' \dots(16),$$

where $F' = F(f, g, h)$. Since F is a homogeneous function it follows that

$$F(\alpha+f, \beta+g, \gamma+h) = F(\alpha, \beta, \gamma) + F(f, g, h) + P,$$

where P consists of a series of products into which at least one of the quantities α, β, γ enters into combination with at least one of the quantities f, g, h.

Since $\Delta^n F$ does not contain α, β, γ it follows that

$$F(f, g, h) = \frac{1}{n!}\Delta^n F(\alpha, \beta, \gamma);$$

similarly $\qquad F(\alpha, \beta, \gamma) = \frac{1}{n!}\Delta'^n F(f, g, h),$

whilst the sum of the remaining terms of (15) and (16) are each equal to P. Now in (15)

$$\frac{1}{p!}\Delta^p F(\alpha,\,\beta,\,\gamma)$$

is the portion of P which is a homogeneous function of $(f,\,g,\,h)$ of degree p; whilst in (16)

$$\frac{1}{(n-p)!}\Delta'^{n-p} F(f,\,g,\,h)$$

is the portion of P which is a similar function of $(f,\,g,\,h)$; whence the two expressions are equal.

24. We have shown in § 22 that the equation of the tangent at any point $(\xi,\,\eta,\,\zeta)$ on a curve is

$$\alpha\frac{dF}{d\xi}+\beta\frac{dF}{d\eta}+\gamma\frac{dF}{d\zeta}=0\ldots\ldots\ldots\ldots\ldots(17),$$

but if the tangent passes through $(f,\,g,\,h)$

$$f\frac{dF}{d\xi}+g\frac{dF}{d\eta}+h\frac{dF}{d\zeta}=0\ldots\ldots\ldots\ldots\ldots(18).$$

Hence the curve

$$\Delta F(\alpha,\,\beta,\,\gamma)=0\ldots\ldots\ldots\ldots\ldots\ldots\ldots(19)$$

passes through the points of contact of all the tangents drawn from the point $(f,\,g,\,h)$ to the curve. This curve is called the *first polar* of $(f,\,g,\,h)$.

If F be of the nth degree, ΔF is of the $(n-1)$th degree; whence a curve and its first polar intersect in $n(n-1)$ points. It therefore follows that from any point *not* on a curve, the maximum number of tangents that can be drawn to the curve is $n(n-1)$; hence not more than six tangents can be drawn to a cubic, nor twelve to a quartic. We shall, however, prove hereafter that when a curve has multiple points the number of tangents is reduced, and that the *class* of every autotomic curve of the nth degree is less than $n(n-1)$.

25. *From any point on a curve, not more than $(n+1)(n-2)$ tangents can be drawn to the curve exclusive of the tangent at the point itself.*

Let O be a point in the neighbourhood of a curve; draw the tangents OP, OQ touching the curve at points near O. Then, excluding OP, OQ not more than $n(n-1)-2$ or $(n+1)(n-2)$ tangents can be drawn from O; but if O moves up to coincidence with P, the two tangents OP, OQ coincide; hence excluding the tangent at P, not more than $(n+1)(n-2)$ tangents can be drawn from P.

26.   *From a point of inflexion, not more than $n(n-1)-3$ tangents can be drawn to a curve.*

At a point of inflexion P the curve cuts its tangent, and the latter has a contact of the second order with the curve.   From a point O near P, draw three tangents OQ, OQ_1, OQ_2, touching the curve at points near O.   Then two of the points of contact will lie on the same side of the tangent at P that O does, whilst the third one will lie on the opposite side.   But when O moves up to coincidence with P all three tangents will coincide with the tangent at P; hence the number of remaining tangents that can be drawn from P to the curve is $n(n-1)-3$.

27.   *From a node, not more than $n(n-1)-4$ tangents can be drawn to a curve.*

Let O be a point on the curve near the node; then we have shown in § 25 that $(n+1)(n-2)$ tangents can be drawn to the curve from O.   But two of these tangents will touch the branch which does *not* pass through O at two points P and Q which are near the node.   Hence when O coincides with the node, these two tangents will coincide with the other nodal tangent, and therefore not more than $(n+1)(n-2)-2=n(n-1)-4$ tangents can be drawn from the node.

28.   *From a cusp, not more than $n(n-1)-3$ tangents can be drawn to a curve.*

Let O be a point on the curve near a cusp; then only one tangent can be drawn from O to touch the other branch in the neighbourhood of the cusp, and when O coincides with the cusp, this tangent coincides with the cuspidal tangent.   Hence the number of tangents which can be drawn from a cusp is

$$(n+1)(n-2)-1=n(n-1)-3.$$

The last four propositions may be stated in a somewhat different form.   If m be the class of a curve, the number of tangents which can be drawn from any point O which is not on the curve is equal to m; and the preceding results show that if O is a node the number of tangents is equal to $m - 4$; if O is a cusp or a point of inflexion, the number is $m - 3$; whilst if O is an ordinary point on the curve, the number is $m - 2$.

29. The equation

$$\Delta^p F(\alpha, \beta, \gamma) = 0 \quad \dots\dots\dots\dots\dots\dots(20)$$

is called the pth polar of the curve with respect to (f, g, h), and is a curve of degree $n - p$. Also by § 23, the pth polar may be written in the form

$$\Delta'^{n-p} F(f, g, h) = 0\dots\dots\dots\dots\dots\dots(21).$$

The $(n - 1)$th polar is therefore a straight line, which is called the *polar line*; whilst the $(n - 2)$th polar is a conic, which is called the *polar conic*. The equations of the polar line and polar conic are

$$\Delta' F' = 0, \text{ and } \Delta'^2 F' = 0 \quad \dots\dots\dots\dots\dots(22).$$

If one of the vertices, say A, of the triangle of reference be taken as the pole, $g = h = 0$, and the pth polar assumes the simple form

$$\frac{d^p F}{d\alpha^p} = 0 \quad \dots\dots\dots\dots\dots\dots\dots (23).$$

By means of (19) of § 24, it can be shown that when a curve is expressed in terms of Cartesian coordinates, the first polar of (f, g) is

$$f \frac{dF}{dx} + g \frac{dF}{dy} + u_{n-1} + 2u_{n-2} + \dots nu_0 = 0\dots\dots\dots(24).$$

30. *The locus of all points, whose polar lines pass through a fixed point, is the first polar of that point.*

Let (f, g, h) be the fixed point; (ξ, η, ζ) any other point. The equation of the polar line of (ξ, η, ζ) is

$$\alpha \frac{dF}{d\xi} + \beta \frac{dF}{d\eta} + \gamma \frac{dF}{d\zeta} = 0 ;$$

but if this pass through the point (f, g, h)

$$f \frac{dF}{d\xi} + g \frac{dF}{d\eta} + h \frac{dF}{d\zeta} = 0,$$

which shows that the locus of (ξ, η, ζ) is the curve $\Delta F = 0$, which is the first polar of (f, g, h).

2—2

In the same way it can be shown that the locus of points, whose polar conics pass through a fixed point, is the second polar of that point, and so on.

31. *The first polars of every point on a straight line pass through the pole of that line.*

Let (f, g, h) be the pole, then the equation of the polar line is

$$\alpha \frac{dF}{df} + \beta \frac{dF}{dg} + \gamma \frac{dF}{dh} = 0 ;$$

hence if (ξ, η, ζ) be any point on this line,

$$\xi \frac{dF}{df} + \eta \frac{dF}{dg} + \zeta \frac{dF}{dh} = 0,$$

which shows that the first polar of (ξ, η, ζ) passes through (f, g, h).

32. *Every straight line has $(n-1)^2$ poles.*

Let P and Q be any two points on a straight line, O its pole. Then by the preceding proposition, the point O lies on the first polars of the curve with respect to P and Q; but these two polars being of the $(n-1)$th degree intersect in $(n-1)^2$ points; hence there are $(n-1)^2$ points which have the same polar line.

33. *The polar line of every point on a curve is the tangent at that point.*

The equation of the polar line of (f, g, h) is $\Delta' F' = 0$; and if (f, g, h) lie on the curve, this is the equation of the tangent at that point.

34. *Every polar of a point on the curve touches the curve at that point.*

If U_p be the pth polar of (f, g, h)

$$U_p = \Delta^p F = 0 \ldots\ldots\ldots\ldots\ldots\ldots(25),$$

which obviously passes through (f, g, h). The equation of the tangent to U_p at (f, g, h) is $\Delta' U_p' = 0$; but since U_p is a ternary quantic of degree $n - p$, this may be written $\Delta^{n-p-1} U_p = 0$ by § 23. Whence substituting the value of U_p from (25), the equation of the tangent to U_p becomes $\Delta^{n-1} F = 0$, which by § 23 is the same thing as $\Delta' F' = 0$, which is the equation of the tangent to F at the point (f, g, h).

35. *The first polar of any point passes through every double point on a curve.*

By § 22, the coordinates of a double point satisfy the equations $dF/d\alpha = 0$, $dF/d\beta = 0$, $dF/d\gamma = 0$, which obviously satisfy the equation $\Delta F = 0$.

36. In § 9 we have defined the *Hessian* of a quantic; we shall now proceed to investigate some of the properties of the curve obtained by equating to zero the Hessian of a ternary quantic, which we shall denote by $H(\alpha, \beta, \gamma) = 0$.

The Hessian of a curve is the locus of the points whose polar conics break up into two straight lines.

The equation of the polar conic is $\Delta'^2 F' = 0$. Let $A = d^2F/df^2$, $F = d^2F/dgdh$ &c. &c., then if the polar conic be written out at full length it becomes

$$A\alpha^2 + B\beta^2 + C\gamma^2 + 2F\beta\gamma + 2G\gamma\alpha + 2H\alpha\beta = 0.$$

The condition that this should break up into two straight lines is that its discriminant should vanish; and the discriminant of the conic is obviously the Hessian of $F(f, g, h)$. Hence

$$H(f, g, h) = 0,$$

and therefore the point (f, g, h) lies on the curve $H(\alpha, \beta, \gamma) = 0$.

37. *The Hessian passes through every double point.*

The coordinates (f, g, h) of a double point satisfy the equations $dF/df = 0$ &c.; and therefore by Euler's theorem

$$Af + Hg + Gh = 0,$$
$$Hf + Bg + Fh = 0,$$
$$Gf + Fg + Ch = 0,$$

which shows that the Hessian $H(f, g, h) = 0$, and therefore the double point lies on the curve $H(\alpha, \beta, \gamma) = 0$.

38. *If the first polar of a point A has a double point at B, then the polar conic of B has a double point at A.*

Let (f, g, h) and (ξ, η, ζ) be the coordinates of A and B. The condition that the first polar of A should have a double point is that the differential coefficients of ΔF should vanish at B. Hence if $A...$, $F...$ denote the second differential coefficients of $F(\xi, \eta, \zeta)$, we must have

$$\left.\begin{array}{l} Af + Hg + Gh = 0 \\ Hf + Bg + Fh = 0 \\ Gf + Fg + Ch = 0 \end{array}\right\} \dots\dots\dots\dots\dots(26),$$

which requires that $H(\xi, \eta, \zeta) = 0$. This shows that if the first polar of a curve has a double point at B, then B must lie on the Hessian; and therefore by § 36, the polar conic of the double point B must break up into two intersecting straight lines. The polar conic of B is

$$A\alpha^2 + B\beta^2 + C\gamma^2 + 2F\beta\gamma + 2G\gamma\alpha + 2H\alpha\beta = 0,$$

and the double point, which is the point of intersection of the two straight lines constituting the conic, is determined by the equations

$$A\alpha + H\beta + G\gamma = 0 \ \&c. \ \&c.,$$

which by (26) are obviously satisfied by (f, g, h).

39. Equations (26) give relations between the coordinates of the points A and B; and if we eliminate (ξ, η, ζ) we shall obtain the locus of A, which is called the *Steinerian* after the German mathematician Steiner. The Steinerian is the locus of the points of intersection of each pair of straight lines which is the polar conic of points on the Hessian.

40. *Every curve of the nth degree has n real or imaginary asymptotes.*

Since an asymptote touches the curve at infinity, it follows that the asymptotes are the tangents at the points where the line at infinity cuts the curve, and there are consequently n asymptotes.

A more analytical proof is furnished by the method for finding asymptotes explained in books on the Differential Calculus. This method consists in substituting $\mu x + \beta$ for y in the Cartesian equation of the curve, and equating the coefficients of x^n and x^{n-1} to zero, which furnishes two equations for determining μ and β. Since the equation for μ is in general of the nth degree, n real or imaginary values of μ exist.

On the General Equation in Trilinear Coordinates.

41.   The general equation of a curve of the nth degree may be written in the form

$$F(\alpha, \beta, \gamma) = u_0\alpha^n + u_1\alpha^{n-1} + u_2\alpha^{n-2} + \dots u_n = 0 \dots\dots(27),$$

where u_n is a binary quantic in β and γ.   The equation may also be written in two similar forms by interchanging the letters α, β and γ.

If the curve pass through the vertex A of the triangle of reference, (27) must be satisfied by $\beta = \gamma = 0$, which requires that $u_0 = 0$.   *Hence if a curve pass through the angular points of the triangle of reference, the terms involving the nth powers of α, β, γ are absent.*

If, in addition, we seek the points where the line $u_1 = 0$ cuts the curve, we find by eliminating γ that the resulting equation contains β^2 as a factor, which shows that the line $\beta = 0$ or CA cuts the curve at a point where u_1 has a contact of the first order with it.   From this it follows that *if a curve pass through the angular points of the triangle of reference the coefficients of the $(n-1)$th powers of α, β and γ equated to zero are the tangents at these points.*

If the point A be a double point, u_1 as well as u_0 must be zero; and $u_2 = 0$ is the equation of the tangents at A.

If therefore the angular points of the triangle of reference are double points, the coefficients of the $(n-2)$th powers of α, β, γ are the tangents at the double points.

If A be a point of inflexion, the tangent at A must meet the curve in three coincident points.   If therefore in (27) we put $u_1 = 0$ and eliminate γ, the resulting equation must contain β^3 as a factor.   This requires that $u_2 = u_1 v_1$, and (27) becomes

$$u_1\alpha^{n-1} + u_1 v_1\alpha^{n-2} + u_3\alpha^{n-3} + \dots u_n = 0 \dots\dots\dots (28).$$

The last result enables us to prove the following important proposition.

42.   *The points of inflexion are the points of intersection of a curve and its Hessian, and their number cannot exceed $3n(n-2)$.*

By § 29, the polar conic of A is

$$\frac{d^{n-2}F}{d\alpha^{n-2}} = 0,$$

whence if A be a point of inflexion, the polar conic is

$$u_1 \{(n-1)\,\alpha + v_1\} = 0 \quad\dots\dots\dots\dots\dots(29),$$

from which it appears that the polar conic of a point of inflexion breaks up into two straight lines, one of which is the tangent $u_1 = 0$, whilst the other is the line $(n-1)\,\alpha + v_1 = 0$. Hence every point of inflexion is a point on the Hessian. Also since the degree of the Hessian is $3(n-2)$, the number of points of inflexion cannot exceed $3n(n-2)$.

If in (27) all the coefficients up to and including u_{k-1} are zero, the vertex A is a multiple point of order k; and the equation $u_k = 0$ determines the k tangents to the curve at A.

43. *If a curve has a multiple point of order k, that point will be a multiple point of order $k-1$ on the first polar, of order $k-2$ on the second, and so on.*

Let A be the multiple point and B the pole. Then the equation of the curve is of the form

$$u_k \alpha^{n-k} + u_{k+1} \alpha^{n-k-1} + \dots u_n = 0 \quad\dots\dots\dots\dots(30),$$

and the first polar of B is

$$\frac{du_k}{d\beta}\,\alpha^{n-k} + \frac{du_{k+1}}{d\beta}\,\alpha^{n-k-1} + \dots \frac{du_n}{d\beta} = 0,$$

and since $du_k/d\beta$ is a binary quantic of degree $k-1$, it follows that A is a multiple point of order $k-1$ on the first polar.

44. *If two tangents at a multiple point coincide, the coincident tangent touches the first polar of every point.*

The equation $u_k = 0$ gives the k tangents at the multiple point A; but if two of them coincide, we must have

$$u_k = (\mu\beta + \nu\gamma)^2\, v_{k-2}.$$

Now the coefficient of α^{n-k} in the first polar of B is

$$(\mu\beta + \nu\gamma)\{2\mu v_{k-2} + (\mu\beta + \nu\gamma)\,dv_{k-2}/d\beta\},$$

which equated to zero gives the tangents at A to the first polar; hence the line $\mu\beta + \nu\gamma = 0$ touches both curves.

Putting $k = 2$, it follows that the tangent at a cusp touches the first polar of every point.

45. *A multiple point of order k on a curve is a multiple point of order $3k - 4$ on the Hessian.*

The equation of a curve having a multiple point at A is given by (30), and if $A = d^2F/d\alpha^2 \ldots$, $F = d^2F/d\beta d\gamma \ldots$, the equation of the Hessian is

$$ABC + 2FGH - AF^2 - BG^2 - CH^2 = 0\ldots\ldots\ldots(31),$$

which is of degree $3n - 6$. Now the degrees of α, β, γ in the different terms are shown in the following table:

	A	B	C	F	G	H
α	$n - k - 2,$	$n - k,$	$n - k,$	$n - k,$	$n - k - 1,$	$n - k - 1,$
β	k ,	$k - 2,$	k ,	k ,	k ,	$k - 1$,
γ	k ,	k ,	$k - 2,$	$k - 2,$	$k - 1$,	k .

From this table it appears that the highest power of α is of degree $3n - 3k - 2$, and that its coefficient is a binary quantic in β and γ of degree $3k - 4$. Hence A is a multiple point on the Hessian of order $3k - 4$.

46. *Every tangent at a multiple point on a curve is a tangent to the Hessian at that point.*

Let the line $\beta = 0$ coincide with *any* tangent through A to the curve; then u_k must contain β as a factor and must therefore be equal to βv_{k-1}. But on referring to the table we see that the highest powers of α in A, C and G must contain β as a factor, and since every term of the Hessian must contain A, C or G, the coefficient of the highest power of α in the Hessian contains β as a factor and therefore this line is the tangent at the point A to the Hessian.

Putting $k = 2$, it follows that every double point on a curve is a double point on the Hessian, and that the tangents at the double point are common to the curve and its Hessian.

Singularities at Infinity.

47. In § 41 we investigated the conditions that a curve should have a double point or a point of inflexion at a finite distance from certain lines of reference; but it frequently happens that a curve has singularities at infinity, and we shall now explain a method by which such singularities may be determined.

Let ABC be the triangle of reference, and let $AB'C'$ be a subsidiary triangle of reference such that the base $B'C'$ cuts the lines AB, AC in B' and C'. Let (α, β, γ) and (α', β, γ) be the trilinear coordinates of a point referred to the two triangles, where $\alpha' = 0$ is the equation of $B'C'$ referred to ABC, and consequently α' is a linear function of α, β, γ.

The equation of a curve having any proposed singularity at B' can be at once written down whenever the nature of the singularity is known. If, however, $B'C'$ be supposed to move off to infinity, $B'C'$ will become the line at infinity, and its equation referred to ABC will be $I = 0$, where

$$I = a\alpha + b\beta + c\gamma \, ;$$

consequently the trilinear equation of a curve having any proposed singularity at infinity upon the line AB may be obtained by first writing down the trilinear equation of a curve having the proposed singularity at B, and then changing α into I.

The general equation of a curve having a double point at B is

$$\beta^{n-2} u_2 + \beta^{n-3} u_3 + \dots u_n = 0 \dots\dots\dots\dots(32),$$

where u_n is a binary quantic in α and γ. Hence the general equation of a curve having a double point at infinity on the line AB is of the same form as (32), where u_n is a binary quantic in I and γ.

48. *To find the equation of a curve having a double point at infinity on the axis of x.*

Let the triangle of reference have a right angle at A, and let AB and AC be the axes of x and y. Then the trilinear equation of a curve having a double point at B is given by (32). Let

$$u_2 = \lambda \alpha^2 + 2\mu\alpha\gamma + \nu\gamma^2 \, ;$$

then when B moves off to infinity, we must write

$$\alpha = I, \quad \beta = x, \quad \gamma = y \dots\dots\dots\dots\dots(33),$$

where I is constant, whence (32) becomes

$$x^{n-2}(\lambda I^2 + 2\mu I y + \nu y^2) + x^{n-3} U_3 + \dots U_n = 0 \dots\dots(34),$$

where U_n is a polynomial of the nth degree in y. Equation (34) is the general equation in Cartesian coordinates of a curve which has a double point at infinity lying on the axis of x. The equation of the tangents at the double point is

$$\lambda I^2 + 2\mu I y + \nu y^2 = 0,$$

and the latter will be a node, a cusp or a conjugate point according as $\mu^2 >$ or $=$ or $< \lambda\nu$.

When $\nu = 0$, the line at infinity is one of the tangents at the double point; and when $\mu = \nu = 0$, the double point is a cusp and the line at infinity is the cuspidal tangent.

49. *To find the equation of a curve having a point of inflexion at infinity on the axis of x.*

The general equation of a curve having a point of inflexion at B is

$$\beta^{n-2}u_1(p\alpha + q\beta + r\gamma) + \beta^{n-3}u_3 + \dots u_n = 0 \dots\dots(35);$$

whence if B is at infinity, the trilinear equation is found by writing I for α; whilst the Cartesian equation is found as in the last article by substituting the values of α, β, γ from (33). Whence if $u_1 = \lambda\alpha + \nu\gamma$, the required equation is

$$x^{n-2}(\lambda I + \nu y)(pI + qx + ry) + x^{n-3}U_3 + \dots U_n = 0\dots(36).$$

The equation of the inflexional tangent is

$$\lambda I + \nu y = 0 \dots\dots\dots\dots\dots\dots\dots(37),$$

and is therefore parallel to the axis of x, excepting in the case in which $\nu = 0$, when it becomes the line at infinity.

50. *To find the condition that the line at infinity should touch the curve.*

If the line $\alpha = 0$ is the tangent at C, the equation of the curve is

$$\gamma^{n-1}\alpha + \gamma^{n-2}u_2 + \dots u_n = 0\dots\dots\dots\dots\dots(38),$$

where u_n is a binary quantic in α and β. Let

$$\alpha = I, \quad \beta = y, \quad \gamma = ax + by,$$

then (38) becomes

$$(ax + by)^{n-1} + (ax + by)^{n-2}U_2 + \dots U_n = 0,$$

where U_n is a polynomial in y. The axis of x joins the origin with the point of contact of the line at infinity with the curve.

By proceeding in a similar manner, we can find the Cartesian equation of a curve with which the line at infinity has a higher contact than the first.

Imaginary Singularities.

51. It frequently happens that a curve has imaginary singularities. Thus in Chapter V. it will be shown that every anautotomic cubic has six imaginary points of inflexion, whilst a quartic may have a pair of imaginary nodes or cusps; but in order that a curve may be real, it is necessary that the number of imaginary singularities of any proposed kind shall be even. We shall now explain a method for determining the conditions for these singularities.

Let ABC be the triangle of reference, and let us construct a subsidiary triangle of reference by taking any two points B', C' on BC. Let (α, β, γ) and $(\alpha, \beta', \gamma')$ be the trilinear coordinates of a point referred to ABC and $AB'C'$. Then $\beta' = 0$, $\gamma' = 0$ will be the equations of AC', AB' referred to ABC, and will be linear functions of β and γ.

Let there be two singularities of the same kind at B' and C', and write down the trilinear equation of a curve referred to $AB'C'$ having these singularities at B', C'. If the singularities are imaginary, B' and C' will be imaginary points, and the lines AB', AC' will also be imaginary; but in order that the curve may be real, it is necessary that AB', AC' should be a pair of conjugate imaginary lines, and their equations must accordingly be of the form $\beta + \iota k\gamma = 0$ and $\beta - \iota k\gamma = 0$, where k is a real constant. We must therefore substitute these values of β', γ' in the equation of the curve, and replace the imaginary constants by new real constants, and the resulting equation will represent a real curve having a pair of conjugate imaginary singularities on the line BC or $\alpha = 0$.

The Cartesian equation of the curve may be obtained by writing

$$\beta = x, \quad \gamma = y, \quad \alpha = Ax + By + C;$$

and the resulting equation will represent a curve having a pair of imaginary singularities of the proposed kind on the line

$$Ax + By + C = 0.$$

When the imaginary singularities are at infinity, we must proceed as before, but write I for α, where $I = 0$ is the line at infinity.

The most interesting case of imaginary singularities at infinity occurs when the singularities are situated at the circular points at infinity; but the discussion of this question must be postponed to a subsequent chapter. We shall merely observe that the Cartesian equation of a curve having a pair of singularities at these points may be obtained by first writing down the trilinear equation of a curve which has the proposed singularities at B and C, and then writing

$$\alpha = \text{const.}, \quad \beta = x + \iota y, \quad \gamma = x - \iota y.$$

CHAPTER III.

TANGENTIAL COORDINATES.

52. IN the Cartesian or the trilinear system of coordinates, a curve is defined as the locus of a point which moves in a prescribed manner. This condition leads to a functional relation between the coordinates of the moving point, which is called the equation of the curve.

In the tangential system, a curve is defined as the envelope of a line which moves in a prescribed manner. Since the position of any straight line is completely determined by means of two independent quantities, the condition that the line should move in the prescribed manner involves a relation between these quantities, which is called the *tangential* equation of the curve.

The system of tangential coordinates which we shall now explain was invented by the late Dr Booth[1] and is sometimes called the Boothian system. Let ξ and η be the reciprocals of the intercepts which a straight line cuts off from the axes; then the equation of the line is

$$x\xi + y\eta = 1 \dots\dots\dots\dots\dots\dots\dots(1),$$

and if this line envelopes a curve, a relation must exist between ξ and η of the form

$$F(\xi, \eta) = 0 \dots\dots\dots\dots\dots\dots\dots(2),$$

which is the tangential equation of the curve.

53. *To find the tangential equation of a curve whose Cartesian equation is given.*

Let the Cartesian equation of the curve be

$$u_n + u_{n-1} + \dots u_1 + u_0 = 0 \dots\dots\dots\dots\dots(3),$$

---

[1] *A Treatise on some New Geometrical Methods.*

where u_n is a binary quantic in x and y.   By (1) this equation may be made homogeneous in x and y, by multiplying each term by the appropriate power of $x\xi + y\eta$, in which case it takes the form

$$u_n + (x\xi + y\eta)\, u_{n-1} + \ldots (x\xi + y\eta)^n\, u_0 = 0 \ldots\ldots\ldots(4).$$

If (4) be divided by x^n, the resulting equation determines n values of $\tan\theta$, where θ is the vectorial angle of the n points in which (1) cuts (3).   If however (1) touches (3), two of the roots of (4) must be equal, the condition for which is that the discriminant of (4) should vanish.   This gives a relation of the form

$$\Delta\,(\xi,\ \eta) = 0 \ldots\ldots\ldots\ldots\ldots\ldots\ldots\ldots\ldots(5),$$

where Δ is the discriminant, which is the required tangential equation.

The discriminants of a binary quadric, cubic and quartic have been given in § 4 ; hence the tangential equation of any conic, cubic or quartic can be obtained by substituting the values of the coefficients of powers of x and y from (4) in the discriminants.

54.   *To find the Cartesian equation of a curve whose tangential equation is given.*

Let the tangential equation be

$$v_n + v_{n-1} + \ldots v_1 + v_0 = 0 \ldots\ldots\ldots\ldots\ldots\ldots\ldots(6),$$

where v_n is a binary quantic in ξ and η.   Make (6) homogeneous in ξ and η by multiplying each term by the appropriate power of $x\xi + y\eta$, and we obtain

$$v_n + (x\xi + y\eta)\, v_{n-1} + \ldots (x\xi + y\eta)^n\, v_0 = 0 \ldots\ldots\ldots(7).$$

Now if ψ be the angle which any tangent drawn from the point (x, y) to the curve makes with the axis of x, $\tan\psi = -\,\xi/\eta$; hence if (7) be divided by η^n, the resulting equation determines the n values of ψ corresponding to the n tangents which can be drawn from (x, y) to the curve.   If, however, the point (x, y) lies on the curve, two of the values of $\tan\psi$ must be equal, the condition for which is that the discriminant of (7) must vanish.   This gives a relation of the form

$$\Delta\,(x,\ y) = 0 \ldots\ldots\ldots\ldots\ldots\ldots\ldots\ldots\ldots(8),$$

which is the Cartesian equation of the curve.

From the last two articles we obtain the following propositions:—

(i) *If $f(x, y) = 0$ is the Cartesian equation of the curve whose tangential equation is $F(\xi, \eta) = 0$, then $f(\xi, \eta) = 0$ is the tangential equation of the curve whose Cartesian equation is*

$$F(x, y) = 0.$$

(ii) *The class of a curve is the same as the degree of its tangential equation.*

55. In practice, the most convenient method of finding the tangential equation of a curve is to write down the equation of the tangent at any point (x, y), which gives the values of (ξ, η) in terms of x and y, and then to eliminate the two latter quantities by means of the equation of the curve. We shall apply this method to find the tangential equation of the curve

$$(x/a)^n + (y/b)^n = 1 \dots\dots\dots\dots\dots(9).$$

The equation of the tangent at (x, y) is

$$\frac{X x^{n-1}}{a^n} + \frac{Y y^{n-1}}{b^n} = 1,$$

where (X, Y) are current coordinates; whence

$$\xi = x^{n-1}/a^n, \quad \eta = y^{n-1}/b^n,$$

and the tangential equation is

$$(a\xi)^{\frac{n}{n-1}} + (b\eta)^{\frac{n}{n-1}} = 1 \dots\dots\dots\dots(10).$$

Equations of curves can also be transformed from Cartesian to tangential coordinates and *vice versâ* by the methods explained in books on the Differential Calculus for finding the envelope of a line. Should, however, a troublesome elimination be necessary, the discriminant may be used with advantage.

56. We must now determine the geometrical meaning of the different terms of a tangential equation.

The equations $\xi = a$, $\eta = b$ represent a line which cuts off from the axes intercepts equal to a^{-1}, b^{-1}; and the equations $\xi = 0$, $\eta = 0$ represent the line at infinity.

The equation $A\xi + B\eta = C$ represents a point whose Cartesian equations are $x = A/C$, $y = B/C$. If $C = 0$, x and y are infinite, and therefore the equation $A\xi + B\eta = 0$, where A and B are any constants, represents a point at infinity.

The equation $A\xi^2 + B\eta^2 = C$ represents a central conic ; for if in (9) and (10) we put $n = 2$, equation (9) represents a conic, whilst (10) is of the preceding form.

The equation $v_2 + v_1 + v_0 = 0$ is the general tangential equation of a conic, since it represents a curve of the second class and conics are the only curves of this class.

The equation $v_3 + v_2 + v_1 + v_0 = 0$ is the general equation of a curve of the third class, and we shall show in Chapter VII. that these curves may be sextics, quartics or cubics.

If $f(\xi, \eta) = 0$ and $F(\xi, \eta) = 0$ be the tangential equations of two curves, the solution of these equations regarded as a pair of simultaneous equations determines the common tangents to the two curves. Hence two curves of the mth and nth classes have mn real or imaginary common tangents. If a pair of roots are equal, two of the common tangents coincide, and the curves touch one another.

Reciprocal Polars.

57.   *If $F(\xi, \eta) = 0$ be the tangential equation of a curve, the Cartesian equation of its reciprocal polar is $F(x/k^2, y/k^2) = 0$.*

Let the tangent at any point of a curve cut the axes in A and B; draw OY perpendicular to AB, and produce it to Q so that $OY \cdot OQ = k^2$. Then the locus of Y is the pedal, and the locus of Q (which is the inverse of the pedal) is the reciprocal polar of the curve with respect to the origin O.

Now, if $YOA = \theta$,

$$OA \cos \theta = OY = k^2/OQ,$$

whence if (x, y) be the coordinates of Q,

$$x = k^2\xi, \quad y = k^2\eta,$$

and the equation of the locus of Q is $F(x/k^2, y/k^2) = 0$.

If therefore we prove any theorem with respect to a curve of given degree, the corresponding property of a curve of the same class can be obtained by reciprocation.

58.   Before proceeding further, we shall state two well known geometrical propositions.

I. *Let OY be the perpendicular on to the tangent at any point P of a curve, and let OZ be the perpendicular from O on to the tangent at Y to the locus of Y; then the angle $OPY = OYZ$; and $OP \cdot OZ = OY^2$.*

II. *Let OP be produced to Q so that $OP \cdot OQ = k^2$, where k is a constant. Let the tangent at Q to the locus of Q meet the tangent at P in T. Then the angle $TPQ = TQP$.*

The locus of Y is the first positive pedal, and the locus of Q is the inverse of the original curve. Also the reciprocal polar is the inverse of the pedal. We can now prove that:—

59. *A node corresponds to a double tangent on the reciprocal polar, and vice versâ.*

Let NY, NY' be the tangents at a node N; from the origin O draw OY, OY' perpendicular to NY, NY', and produce them to Q, Q' so that

$$OY \cdot OQ = OY' \cdot OQ' = k^2 \quad \ldots\ldots\ldots\ldots(11).$$

Join QQ', YY' and draw TY such that the angle $TYQ = TQY$.

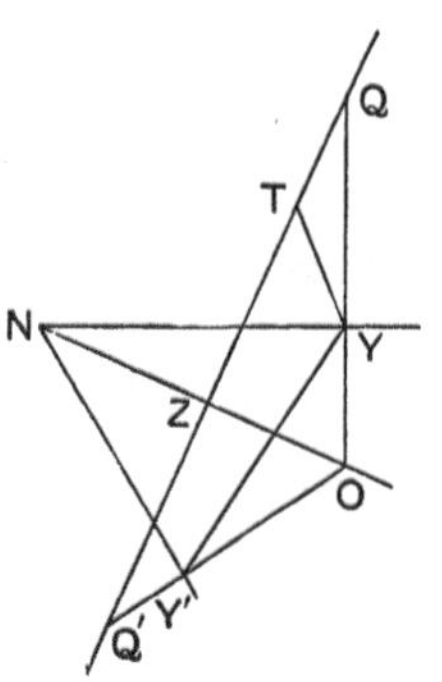

From (11) it follows that a circle can be described through $QYY'Q'$; also a circle can be described through $OYNY'$; whence

$$Q'QY = YY'O = ONY,$$

accordingly a circle can be described through $NZYQ$, and therefore the angle NZQ is a right angle. Whence

$$OZ \cdot ON = OY \cdot OQ = k^2 \text{ and } TYQ = TQY = ONY,$$

and therefore TY is the tangent at Y to the pedal, and TQ is the tangent at Q to the reciprocal polar.

Similarly TQ' is the tangent at Q', and therefore QQ' touches the reciprocal polar at Q and Q'. Also since $OZ \cdot ON = k^2$, QQ' is the polar of N.

Since a conjugate point is a *real* point, its polar is a *real* line; but since the tangents at a conjugate point are imaginary, the double tangent corresponding to a conjugate point touches the reciprocal curve at two *imaginary* points.

60. *A cusp corresponds to a stationary tangent on the reciprocal polar, and vice versâ.*

Let S be any point on the cuspidal tangent near the cusp N. Let SN_1, SN_2 be the tangents from S to the two branches which touch at the cusp; and let Q, Q_1, Q_2 be the three points on the reciprocal polar which correspond to the tangents SN, SN_1, SN_2. Since the three tangents intersect at a point S, the three points Q, Q_1, Q_2 lie on a straight line which is the polar of S; accordingly when S moves up to coincidence with N, the straight line QQ_1Q_2 has a contact of the second order, and is therefore a stationary tangent to the reciprocal polar.

The Line at Infinity.

61. When the equation of a curve is given in Cartesian coordinates, the absolute term can always be got rid of by transferring the origin to a point on the curve; but in tangential equations it is impossible by any change of the origin or the axes to get rid of the absolute term, if it exists, or to introduce one if it does not exist. If in (6) $v_0 = 0$, the equation is satisfied by $\xi = 0$, $\eta = 0$, which are the coordinates of the line at infinity; in which case this line is a tangent to the curve. This will happen whenever the curve is the reciprocal polar of another curve with respect to a point on the latter. For example, the reciprocal polar of a conic with regard to any point not on the curve is a central conic; but if the point lie on the conic the reciprocal polar is a parabola, which touches the line at infinity. When the linear term as well as the absolute term is absent, the line at infinity is a stationary or a double tangent according as the quadratic term is or is not a perfect square. In the former case, the curve is the reciprocal polar of some other curve with respect to a cusp, and in the latter with respect to a node. Moreover the points of contact will be real, imaginary, or coincident, according as the double point is a node, a conjugate point or a cusp. And generally, if v_k is the term of lowest degree

in the tangential equation, the line at infinity is a multiple tangent of order k, and the curve is the reciprocal polar of another curve with respect to a multiple point of the same order.

Multiple Tangents.

62.　We shall now employ equation (4) of § 53 to find the multiple tangents to a curve. This equation determines the vectorial angle of the points in which the straight line

$$x\xi + y\eta = 1 \quad \dots\dots\dots\dots\dots\dots\dots(12)$$

cuts the curve $\Sigma_0^n u_r = 0$, and we shall denote it by

$$F(m) = 0 \dots\dots\dots\dots\dots\dots\dots(13),$$

where $m = \tan \theta$.

(i)　If three of the roots of (13) are equal, (12) has a contact of the second order with the curve. The conditions for this are that the discriminants Δ, Δ' of $F(m)$ and $F'(m)$ should vanish. This leads to two equations of the form $\Delta(\xi, \eta) = 0$, $\Delta'(\xi, \eta) = 0$, which are the tangential equations of the original curve and of a second one, such that every line which has a contact of the second order with the original curve is a tangent to the latter curve.

(ii)　If two pairs of roots of (13) are equal, (12) has a contact of the first order with the curve at two distinct points.

(iii)　If four roots are equal, (12) has a contact of the third order with the curve.

The preceding method does something more than determine the multiple tangents to curves. In the case of a cubic the two nodal tangents, as well as the stationary tangents, have a contact of the second order with the cubic. Hence if the origin is not a node, this method will determine the nodal as well as the stationary tangents. So also in the case of a quartic, every ordinary tangent drawn from a double point to the curve, and also every line joining a pair of double points, has a contact of the first order with the curve at two distinct points; hence this method will not only determine the double tangents, but also the tangents drawn from each double point to the curve, together with the lines joining each pair of double points.

The conditions for the different equalities which can exist between the roots of cubic and quartic equations are given in

§ 7, whence the necessary equations for determining the multiple tangents to these curves are obtained by substituting in these equations the values of the coefficients of powers of x and y in (4). For curves of any given degree, the necessary equations can be obtained from the equalities which must exist between the roots of the corresponding equations in one variable. Thus we may find the conditions that a sextic curve may have (i) a triple tangent, (ii) a double tangent touching the curve at two points of inflexion, (iii) a double tangent touching the curve at a point of undulation and having a contact of the first order at the other point.

63. We shall illustrate this method by finding the double tangents to the symmetrical quartic curve

$$Ax^4 + 2Bx^2y^2 + Cy^4 + ax^2 + by^2 = 0 \ldots\ldots\ldots\ldots(14).$$

This curve has a node at the origin, and if we transform to polar coordinates, it will be found that for every assigned value of θ there are two equal values of r, one of which is positive and the other negative. Hence the quartic is *uninodal*, and it will be shown in Chapter VIII. that its class is ten and the number of double tangents is sixteen.

If $x = e$ is a double tangent, it follows that if e be substituted for x in (14) the two values of y^2 must be equal. This gives the equation

$$(b + 2Be^2)^2 = 4Ce^2 (Ae^2 + a)\ldots\ldots\ldots\ldots\ldots(15),$$

which shows that there are four double tangents parallel to y. In the same way it can be shown that there are four double tangents parallel to x. We have thus accounted for eight double tangents. To find the remainder, we write down the equation for m which is

$$m^4 (C + b\eta^2) + 2m^3 b\xi\eta + m^2 (2B + b\xi^2 + a\eta^2) + 2ma\xi\eta + A + a\xi^2 = 0$$
$$\ldots\ldots\ldots\ldots(16),$$

whence, by (16) of § 7, the equations of condition are

$$\left. \begin{array}{l} a^2\xi^2\eta^2 (C + b\eta^2) = b^2\xi^2\eta^2 (A + a\xi^2) \\ 4b^3\xi^3\eta^3 + a (C + b\eta^2)^2 \xi\eta = 3b\xi\eta (C + b\eta^2)(2B + b\xi^2 + a\eta^2) \end{array} \right\} \ldots\ldots(17).$$

Dividing out by the extraneous factor $\xi\eta$, the first equation is the tangential equation of a central conic, whilst the second represents a curve of the fourth class. These two curves have eight common tangents, which are the remaining double tangents to the quartic.

We must now consider the meaning of the extraneous factor $\xi\eta$. Its existence shows that the equations of condition are satisfied by $\xi = 0$, η arbitrary; or $\eta = 0$, ξ arbitrary. We must therefore go back to (16) and put $\eta = 0$, and determine the conditions that the resulting equation in m should have two pairs of equal roots. This will be found to lead to equation (15), which gives the four double tangents parallel to y. In the same way if we put $\xi = 0$ in (16), we shall obtain the four double tangents parallel to x. Equations (16) and (17) accordingly completely determine the sixteen double tangents.

Pedal Curves. Inversion.

64. The locus of the foot of the perpendicular from any origin O on to the tangent at any point of a curve is called the *first positive pedal* of the curve with respect to the origin.

The pedal of the first pedal with respect to the same origin is called the *second positive pedal* of the original curve, and so on.

The curve, of which the original curve is the first positive pedal, is called the *first negative pedal* of the original curve, and so on.

Since the reciprocal polar of a curve is the inverse of its first positive pedal, it follows that the inverse of the original curve is the reciprocal polar of its first negative pedal.

The polar equation of the pedal of a curve gives a relation of the form

$$p = F(\chi)\dots\dots\dots\dots\dots\dots\dots(18),$$

where p is the perpendicular from the origin on to the tangent to the original curve, and χ is the angle which p makes with a fixed straight line. This equation has been called by Dr Ferrers the *tangential polar equation* of the original curve.

The tangential polar equation is useful in finding the envelope of a line; for if any relation of the form (18) can be recognized as the pedal of some known curve, the envelope of the line is the curve in question.

The inverse of a curve, with respect to any origin O, is found by transferring the Cartesian equation to O, and then writing

k^2x/r^2, k^2y/r^2 for x and y; hence u_n becomes $k^{2n}u_n/r^{2n}$, and the equation of the inverse is

$$k^{2n}u_n + k^{2n-2}r^2u_{n-1} + \dots k^2 r^{2n-2}u_1 + r^{2n}u_0 = 0 \dots\dots(19).$$

The degree of the inverse of a curve of the nth degree is in general $2n$; but if the origin be a multiple point of order k, the degree of the inverse will be $2n - k$. The degree will be still further reduced if u_n, u_{n-1} &c. contain some power of r as a factor.

Since the degree of the reciprocal polar is equal to the class of the curve, the degree of the pedal can be found by inversion.

65.   *To find the Cartesian equation of the pedal of a curve.*

Let $\phi(\xi, \eta) = 0$ be the tangential equation of a curve; let any tangent cut the axes of x and y in A and B; also let (x, y) be the coordinates of Y, the foot of the perpendicular from O on to the tangent AB.   Then if

$$AOY = \theta,$$
$$OY = OA \cos\theta = OB \sin\theta,$$

whence

$$x = \frac{\xi}{\xi^2 + \eta^2}, \qquad y = \frac{\eta}{\xi^2 + \eta^2},$$

and

$$\xi = \frac{x}{x^2 + y^2}, \qquad \eta = \frac{y}{x^2 + y^2}.$$

Hence the Cartesian equation of the pedal is

$$\phi\left\{\frac{x}{x^2 + y^2}, \ \frac{y}{x^2 + y^2}\right\} = 0.$$

66.   *To find the tangential equation of the first negative pedal.*

If $F(x, y) = 0$ be the Cartesian equation of the curve, it follows from the preceding formulae, that the required tangential equation is

$$F\left\{\frac{\xi}{\xi^2 + \eta^2}, \ \frac{\eta}{\xi^2 + \eta^2}\right\} = 0.$$

By means of the preceding results, it may be shown that the Cartesian equation of the first positive pedal of the curve

$$(x/a)^n + (y/b)^n = 1$$

is

$$(x^2 + y^2)^{n-1} = (ax)^{\frac{n}{n-1}} + (by)^{\frac{n}{n-1}},$$

and that the tangential equation of its first negative pedal is

$$(\xi^2 + \eta^2)^n = (\xi/a)^n + (\eta/b)^n.$$

On the Curves $r^n = a^n \cos n\theta$.

67. We have investigated several theorems concerning the important class of curves included in the equation

$$(x/a)^n + (y/b)^n = 1 \, ;$$

we shall now consider the equation

$$r^n = a^n \cos n\theta,$$

which includes many important and well known curves.

By the ordinary formula

$$\tan \phi = r\, d\theta/dr = -\cot n\theta,$$

whence

$$\phi = \tfrac{1}{2}\pi + n\theta.$$

Accordingly if (p, χ) be the coordinates of Y,

$$\chi = \theta + \phi - \tfrac{1}{2}\pi = (n + 1)\,\theta,$$

$$p = r \sin \phi = r \cos n\theta\dots\dots\dots\dots\dots\dots\dots(20),$$

whence

$$p^{\frac{n}{n+1}} = a^{\frac{n}{n+1}} \cos n\chi/(n + 1) \, \dots\dots\dots\dots \, (21).$$

Equation (21) is the pedal of the curve; from which it follows that every pedal is a curve of the same species, and that each successive pedal is obtained from the preceding one by changing n into $n/(n + 1)$. The reciprocal polar is the curve

$$c^{\frac{n}{n+1}} = r^{\frac{n}{n+1}} \cos n\chi/(n + 1) \, \dots\dots\dots\dots \, (22),$$

and is obtained by changing n into $- n/(n + 1)$.

From (20) we obtain

$$a^n p = r^{n+1}\dots\dots\dots\dots\dots\dots\dots\dots\dots(23),$$

which is the p and r equation of the curve. The radius of curvature is

$$\rho = r\,\frac{dr}{dp} = \frac{a^n}{(n + 1)\, r^{n-1}} \, \dots\dots\dots\dots\dots(24).$$

Orthoptic Loci.

68. The *orthoptic* locus of a curve is the locus of the point of intersection of two tangents which cut one another at right angles. If the two tangents are inclined at a *constant* angle, the locus is called the *isoptic* locus.

If $\Sigma_0^m v_r = 0$ be the tangential equation of a curve of the mth class, we have shown, in § 54, that the equation

$$\Sigma_0^m v_r \, (x\xi + y\eta)^{m-r} = 0 \dots\dots\dots\dots\dots\dots(25)$$

determines the angles which the m tangents, which can be drawn to the curve from the point (x, y), make with the axis of x. Hence if we write $\xi/\eta = -\tan \psi = -z$, equation (25) may be written in the form

$$f(z) = 0 \ \dots\dots\dots\dots\dots\dots\dots(26),$$

where f is of degree m.

If two of the tangents are at right angles, two of the roots z_1, z_2 of (26) must be connected together by the equation $z_1 z_2 = -1$. The condition for this is that the eliminant of $f(z) = 0$ and $f(-z^{-1}) = 0$ should vanish, which gives a relation between x and y, which is the orthoptic locus.

When $f(z)$ is a quartic, cubic or quadric function, the values of the eliminants are given in § 8; hence the orthoptic locus of any curve of the fourth or any lower class can be determined.

For a curve of the mth class, the eliminant is of degree $m - 1$ in the coefficients, and the coefficients themselves are in general of degree m in x and y. Hence the degree of the orthoptic locus of a curve of the mth class cannot be greater than $m(m-1)$.

We have shown in § 61 that if the curve touch the line at infinity, the absolute term will not appear in the tangential equation. In this case the coefficients in the eliminant are of degree $(m-1)$, and the orthoptic locus of degree $(m-1)^2$. Thus the orthoptic locus of a central conic is a circle, whilst that of a parabola is a straight line.

If the linear as well as the absolute term is absent, the line at infinity is a double or a stationary tangent. In this case, the coefficients are of degree $m-2$, and the orthoptic locus is of degree $(m-2)(m-1)$. And generally if the line at infinity is a multiple tangent of order k, the degree of the orthoptic locus of a curve of the mth class is $(m-k)(m-1)$.

The Cartesian equation of the evolute of a parabola is $ay^2 = x^3$, and its tangential equation is $4a\xi^3 = 27\eta^2$. Hence the line at infinity is a stationary tangent; and it is shown in books on Conics that the orthoptic locus is a parabola.

The tangential equation of the evolute of an ellipse is

$$a^2/\xi^2 + b^2/\eta^2 = (a^2 - b^2)^2,$$

and therefore the evolute is a curve of the fourth class, and the line at infinity is a double tangent which touches the curve at two imaginary points. Hence the orthoptic locus is a sextic curve, whose equation can be shown to be

$$(a^2 + b^2)(x^2 + y^2)(a^2 y^2 + b^2 x^2)^2 = (a^2 - b^2)^2 (a^2 y^2 - b^2 x^2)^2.$$

The Circular Points at Infinity.

69. It is proved in treatises on Trilinear Coordinates[1] that the equation of every circle can be expressed in the form

$$S + (l\alpha + m\beta + n\gamma)\, I = 0,$$

where S is any *given* circle, and I is the line at infinity. The constants (l, m, n) determine the position of the circle and its radius; whilst the form of this equation shows that all circles pass through the points of intersection of a *given* circle with the line at infinity. These two points, which are imaginary, are called the *circular points at infinity* and are usually denoted by the letters I and J.

If $S = 0$ be the equation of the circle circumscribing the triangle of reference, the circular points are the intersections of $S = 0$, $I = 0$; that is of

$$\beta\gamma \sin A + \gamma\alpha \sin B + \alpha\beta \sin C = 0,$$
$$\alpha \sin A + \beta \sin B + \gamma \sin C = 0.$$

Solving these equations, we obtain

$$\alpha = -\gamma\epsilon^{\mp\iota B}, \quad \beta = -\gamma\epsilon^{\pm\iota A} \quad\ldots\ldots\ldots\ldots(27),$$

which are the trilinear coordinates of the circular points at infinity.

70. *To find the Cartesian equations of the lines joining any point with the circular points at infinity.*

Let $y = mx$ be the equation of any line joining the origin with one of the circular points. The points of intersection of this line with the circle $x^2 + y^2 = a^2$ are given by the equation

$$m^2 + 1 = a^2/x^2 \quad\ldots\ldots\ldots\ldots\ldots\ldots(28).$$

---

[1] Ferrers' *Trilinear Coordinates*, p. 87.

Equation (28) shows that if m were real, the straight line would intersect the circle in two real points at a finite distance from the origin; but if $m = \pm \iota$, the left-hand side of (28) vanishes, which shows that x must be infinite. Hence the two imaginary straight lines

$$x \pm \iota y = 0$$

intersect the circle in two imaginary points at infinity, which are the circular points in question. Both lines are included in the equation $x^2 + y^2 = 0$.

Similarly if (α, β) be the Cartesian coordinates of any other point, the equations of the lines joining (α, β) with the circular points are

$$x - \alpha \pm \iota (y - \beta) = 0,$$

both of which are included in the equation $(x - \alpha)^2 + (y - \beta)^2 = 0$.

71. There is another method of tangential coordinates which is founded on the trilinear system.

Let $$\lambda\alpha + \mu\beta + \nu\gamma = 0$$

be any straight line; then the condition that this line should touch the curve $F(\alpha, \beta, \gamma) = 0$ involves a relation between (λ, μ, ν) of the form $\phi(\lambda, \mu, \nu) = 0$, which is the tangential equation of the curve. All the results in this system may be obtained by the preceding methods by writing

$$\alpha/\gamma = x, \quad \beta/\gamma = y, \quad -\lambda/\nu = \xi, \quad -\mu/\nu = \eta.$$

The tangential equation of the conic

$$l\alpha^2 + m\beta^2 + n\gamma^2 + 2l'\beta\gamma + 2m'\gamma\alpha + 2n'\alpha\beta = 0$$

is

$$(mn - l'^2)\lambda^2 + (nl - m'^2)\mu^2 + (lm - n'^2)\nu^2$$
$$+ 2(m'n' - ll')\mu\nu + 2(n'l' - mm')\nu\lambda + 2(l'm' - nn')\lambda\mu = 0.$$

This result may be obtained by eliminating γ between the equation of the conic and the line $\lambda\alpha + \mu\beta + \nu\gamma = 0$, and expressing the condition that the resulting quadratic in α/β should have equal roots. Moreover if α, β, γ and λ, μ, ν be interchanged, the first equation will represent the tangential equation of the conic whose trilinear equation is the second one.

The condition that the line (λ, μ, ν) should touch the curve $F(\alpha, \beta, \gamma) = 0$ at the point (ξ, η, ζ) is sometimes useful. Since

(α, β, γ) and (ξ, η, ζ) satisfy the equations $\lambda\alpha + \mu\beta + \nu\gamma = 0$ and $a\alpha + b\beta + c\gamma = 2\Delta$, it follows that

$$\lambda (\alpha - \xi) + \mu (\beta - \eta) + \nu (\gamma - \zeta) = 0,$$
$$a (\alpha - \xi) + b (\beta - \eta) + c (\gamma - \zeta) = 0,$$

whence

$$\frac{\alpha - \xi}{\mu c - \nu b} = \frac{\beta - \eta}{\nu a - \lambda c} = \frac{\gamma - \zeta}{\lambda b - \mu a},$$

accordingly by (11) of § 22 the condition is

$$(\mu c - \nu b)\frac{dF}{d\xi} + (\nu a - \lambda c)\frac{dF}{d\eta} + (\lambda b - \mu a)\frac{dF}{d\zeta} = 0.$$

72. The foregoing system of tangential coordinates may be exhibited in a geometrical form. Let the line (λ, μ, ν) cut the sides BC, CA, AB of the triangle of reference in D, E and F; and let p, q, r be the lengths of the perpendiculars from A, B, and C on to it; also let any two of these perpendiculars, say p and q, be considered to have contrary signs when the line cuts AB at a point lying between A and B, and in other cases to have the same sign. Then if F lies between A and B,

$$\frac{q}{p} = -\frac{BF}{AF}.$$

Putting $\gamma = 0$ in the equation of DEF, we obtain

$$\frac{\mu}{\lambda} = -\frac{\alpha}{\beta} = -\frac{BF \sin B}{AF \sin A} = \frac{qb}{pa},$$

whence

$$\frac{\lambda}{pa} = \frac{\mu}{qb} = \frac{\nu}{rc} \quad \dots\dots\dots\dots\dots\dots(29),$$

which shows that λ, μ, ν are proportional to the products of the lengths of each perpendicular into the lengths of the opposite sides. The equation of DEF may now be written

$$pa\alpha + qb\beta + rc\gamma = 0,$$

which shows that the coordinates of the line at infinity are $p = q = r$, or $\lambda/a = \mu/b = \nu/c$.

73. *To find the tangential equation of a circle.*

Let

$$\rho^2 = \lambda^2 + \mu^2 + \nu^2 - 2\mu\nu \cos A - 2\nu\lambda \cos B - 2\lambda\mu \cos C,$$

then it is shown in works on Trilinear Coordinates[1] that if ϖ be the perpendicular from any point (f, g, h) on to the line (λ, μ, ν)

$$\varpi\rho = \lambda f + \mu g + \nu h \dots\dots\dots\dots\dots(30).$$

If the envelope is a circle, ϖ is constant, whence (30) is the tangential equation of a circle of radius ϖ and centre (f, g, h).

When the centre of the circle is at A, $f = 2\Delta/a$, $g = h = 0$, $\varpi = p$; whence (30) becomes $p\rho a = 2\Delta\lambda$, which by (29) is the same thing as

$$p^2 a^2 + q^2 b^2 + r^2 c^2 - 2qrbc \cos A - 2rpca \cos B - 2pqab \cos C = 4\Delta^2$$
$$\dots\dots\dots(31),$$

and gives an identical relation between p, q, and r. Equation (31) consequently shows that the three coordinates of any line satisfy a given relation which is independent of the position of the line —a result which might be anticipated from the fact that two coordinates are sufficient to determine a straight line.

74. The trilinear coordinates (α, β, γ) of a point satisfy the identical relation $a\alpha + b\beta + c\gamma = 2\Delta$; but there are certain exceptional points which satisfy the equation $a\alpha + b\beta + c\gamma = 0$, which is the line at infinity. In the same way it may be anticipated that there are certain exceptional lines which satisfy the equation obtained by putting $\Delta = 0$ in (31). To interpret this result put $\varpi = \infty$ in (30); in which case, since λ, μ, ν and f, g, h are finite, we must have $\rho = 0$. The latter equation apparently represents a circle of infinite radius; but as a matter of fact it represents the circular points at infinity. For when $\Delta = 0$, (31) may be written in the form

$$(\lambda \cos B + \mu \cos A - \nu)^2 + (\lambda \sin B - \mu \sin A)^2 = 0.$$

Resolving the left-hand side into factors, the equation is equivalent to the two linear equations

$$\lambda \epsilon^{\iota B} + \mu \epsilon^{-\iota A} - \nu = 0,$$
$$\lambda \epsilon^{-\iota B} + \mu \epsilon^{\iota A} - \nu = 0,$$

which represent the two points

$$\alpha = -\gamma \epsilon^{\pm \iota B}, \quad \beta = -\gamma \epsilon^{\mp \iota A}$$

which are the circular points at infinity.

[1] Ferrers' *Trilinear Coordinates*, p. 20.

Upon this result Prof. Cayley has founded his theory of the Absolute, which has been developed by Prof. Klein and others; but the subject is beyond the scope of an elementary work[1].

Foci.

75. We shall now explain how the circular points are employed to determine the foci of curves, and shall begin by proving that :—

The lines joining the focus of a conic with either of the circular points at infinity touch the conic.

Let (α, β) be the coordinates of the focus of the ellipse

$$x^2/a^2 + y^2/b^2 = 1.$$

The equation of the line joining (α, β) to one of the circular points is

$$x - \alpha + \iota (y - \beta) = 0 \dots\dots\dots\dots\dots(32).$$

Let

$$\xi = \frac{1}{\alpha + \iota\beta}, \qquad \eta = \frac{\iota}{\alpha + \iota\beta} \dots\dots\dots\dots(33),$$

then if (32) is a tangent to the ellipse, ξ and η must be connected by the equation

$$a^2\xi^2 + b^2\eta^2 = 1.$$

Substituting from (33), we obtain

$$\alpha + \iota\beta = \pm (a^2 - b^2)^{\frac{1}{2}}.$$

If $a > b$, the real values of α and β are given by

$$\alpha = \pm (a^2 - b^2)^{\frac{1}{2}}, \ \beta = 0 \dots\dots\dots\dots\dots(34),$$

whilst the imaginary ones are given by

$$\alpha = 0, \ \beta = \pm \iota (a^2 - b^2)^{\frac{1}{2}}.$$

Equations (34) are the well known equations for determining the real foci of the conic.

When the ellipse degenerates into a circle, $a = b$, and the two real foci coincide with the centre, which is a *double* focus.

[1] Cayley, "A sixth Memoir on Quantics," *Math. Papers*, Vol. II. p. 561. Klein, *Math. Annalen*, Vol. XXXVII.; *Lectures on Nicht-Euclidische Geometrie*, Vol. I. p. 61.

The equation $(x - \alpha)^2 + (y - \beta)^2 = 0$ is sometimes regarded as the equation of the point (α, β), since it is the limiting form of the equation of an indefinitely small circle which coincides with this point; but since in the Cartesian system two equations are required to determine a point, the preferable mode of interpretation is to regard the equation as representing two imaginary straight lines through the point (α, β). If, however, we adopt the former mode of interpretation, a focus may be defined as *an indefinitely small circle which has a double contact with the conic.*

76. The foregoing considerations led Plücker[1] to adopt a generalized conception of the foci of curves of a higher degree than the second, which he defined as *the points of intersection of the tangents drawn to a curve from the circular points at infinity.* Since m tangents can in general be drawn to a curve of the mth class, $2m$ tangents can in general be drawn from the two circular points to the curve. All these tangents are imaginary, and they will intersect in m^2 points; but only m of these points will be real, for if one of the tangents drawn from the circular point I be of the form $A + \iota B = 0$, one of the tangents drawn from J will be of the form $A - \iota B = 0$, whilst all the others will be of the form $C - \iota D = 0$. The first tangent from J will intersect the tangent from I at the real point $A = 0$, $B = 0$; but none of the other tangents from J can intersect the tangent from I in a real point unless $C/A = D/B$, in which case the two tangents $A - \iota B = 0$ and $C - \iota D = 0$ become identical. Hence the real foci of the curve are the m real points of intersection of the tangents drawn from the circular points at infinity to the curve, and their number cannot exceed the class of the curve; but if the curve passes through or has singularities at the circular points, the number of foci must be determined by a special investigation.

77. *If the line at infinity is a multiple tangent of order g, a curve of the mth class cannot have more than $m - g$ real foci.*

Let the tangential equation of the curve be

$$u_m + u_{m-1} + \dots u_g = 0 \dots\dots\dots\dots\dots\dots\dots(35),$$

the form of which shows that the line at infinity is a multiple tangent of order g. If (α, β) be a focus, it follows from § 75 that

---

[1] *Crelle*, Vol. x. p. 84; Cayley, "On Polyzomal Curves," *Trans. Roy. Soc. Edinburgh*, Vol. xxv. pp. 1—110; *Collected Papers*, Vol. vi. p. 515.

their values are determined by substituting $(\alpha + \iota\beta)^{-1}$ and $\iota(\alpha + \iota\beta)^{-1}$ for ξ and η in (35). Hence ξ and η will respectively be of the forms $\rho\epsilon^{\iota\theta}$, $\iota\rho\epsilon^{\iota\theta}$, where $\alpha^2 + \beta^2 = \rho^{-2}$, $\tan\theta = -\beta/\alpha$. Substituting in (35) and putting z for $\rho\epsilon^{\iota\theta}$, we obtain

$$z^{m-g}u'_m + z^{m-g-1}u'_{m-1} + \ldots u'_g = 0 \ldots\ldots\ldots\ldots(36),$$

where u'_m is what u_m becomes when $\xi = 1$, $\eta = \iota$.

Equation (36) determines $m - g$ values of z, all of which are complex; and if $A + \iota B$ be any one of these values, the corresponding values of α and β are given by

$$1 = (A + \iota B)(\alpha + \iota\beta),$$

whence

$$\alpha = \frac{A}{A^2 + B^2}, \quad \beta = -\frac{B}{A^2 + B^2},$$

which determine the $m - g$ values of α and β.

78. *If an anautotomic curve of the mth class passes through the circular points at infinity, the curve has m − 2 real single foci and one real double one, which is the point of intersection of the tangents at the circular points.*

When a curve of the nth degree and mth class passes through one of the circular points at infinity, its equation must be of the form

$$Su_{n-2} + Iu_{n-1} = 0 \ldots\ldots\ldots\ldots\ldots\ldots\ldots(37),$$

where u_n is any ternary quantic in α, β, γ. The form of (37) shows that if a curve passes through one circular point it must pass through the other; hence by § 25 the number of tangents which can be drawn from a circular point, exclusive of the tangent at the point itself, is $m - 2$, which is the number of real single foci. The two tangents at the circular points are the limiting positions of the four tangents which can be drawn from two imaginary points I_1, J_1 in the neighbourhood of each circular point and which respectively touch the curve at four points, two of which are near I and the remaining two near J. The two tangents from I_1 intersect the two tangents from J_1 in four points, two of which are real and two imaginary; but when the points I_1, J_1 move up to coincidence with I and J, the two real points of intersection coincide and form a *double* focus situated at the point of intersection of the tangents at the circular points.

79. *If the circular points are nodes, a binodal curve has $m-4$ real single foci and two real double ones, which are the two real points of intersection of the nodal tangents at the circular points.*

When the circular points are nodes, the number of tangents which can be drawn from I exclusive of the two nodal tangents is $m-4$, which is the number of real single foci. Now any one of the nodal tangents at I intersects the conjugate nodal tangent at J in a real point, whilst its point of intersection with the other nodal tangent at J will be imaginary. By § 78, the real point of intersection of a pair of conjugate nodal tangents is a double focus; and since there are two pairs of conjugate nodal tangents, there will be two real double foci.

80. *If the circular points are cusps, a bicuspidal curve has $m-3$ real single foci and one real triple focus, which is the point of intersection of the cuspidal tangents at the circular points.*

When the circular points are cusps, the number of tangents which can be drawn from I, exclusive of the cuspidal tangent, is $m-3$, which is the number of real single foci. Let I_1, J_1 be two imaginary points in the neighbourhood of I and J. Then from I_1 three tangents can be drawn to the curve which touch it at three points near I; and in like manner three similar tangents can be drawn from J_1. These two systems of three tangents will intersect one another in nine points; but since the tangents are all imaginary, each tangent of the I_1 system will intersect the three tangents of the J_1 system in three points, only one of which can be real; and thus there will be altogether three real and six imaginary points of intersection. But when the points I_1, J_1 respectively move up to coincidence with I and J, the nine points will coincide with the point of intersection of the cuspidal tangents at I and J. Hence this point will be a real triple focus.

It can be shown in the same manner that if the circular points are points of inflexion, the curve has the same number of single foci, and one triple focus which is the point of intersection of the stationary tangents at the circular points.

When the line at infinity is a multiple tangent of order g, and the curve in addition possesses any of the above-mentioned singularities, the number of foci is obtained by changing m into $m-g$ in the preceding results.

In §§ 78, 79 and 80, the enunciation has been restricted to anautotomic, binodal and bicuspidal curves respectively. The reason of this is that every line joining the circular points to a double point has a contact of the first order with the curve at the double point, and may therefore be regarded as satisfying Plücker's definition of a focus, in which no distinction is drawn between contact and tangency. If therefore a curve has δ nodes and κ cusps *exclusive of the circular points*, the class m of the curve must be replaced by $m + 2\delta + 3\kappa$ in the formulae giving the number of real single foci. For example, the limaçon is a quartic curve of the fourth class which has a pair of cusps at the circular points and a node at the origin; hence the curve has one triple focus and three single foci. One of the single foci is an isolated point, whilst the node is a double focus formed by the union of the two other single foci. Now it will be shown hereafter that the limaçon is a special form of the oval of Descartes, which is a quartic of the sixth class having a pair of cusps at the circular points and no other double point. The latter curve has one triple focus and three collinear single foci; and when the curve becomes a limaçon two of the single foci unite at the node, so that the limaçon has one triple, one double and one single focus. Similarly by considering the degeneration of the oval of Descartes into a cardioid, it can be shown that the latter curve has one ordinary triple focus, and a triple focus at the cusp formed by the union of the three ordinary single foci of the oval of Descartes.

81. *If a curve be inverted from any point O, the inverse points of the foci of the original curve are the foci of the inverse curve.*

If S be any circle which has a double contact with a curve at the points P, Q, the inverse of S will be another circle which has a double contact with the inverse curve at the inverse points P', Q'. Now we have shown in § 75 that a focus may be regarded as the centre of an indefinitely small circle which has a double contact with the curve; hence the inverse of a focus is an indefinitely small circle which has a double contact with the inverse curve, and is therefore itself a focus.

In considering the properties of the foci of curves, it has been usual to restrict the discussion to real foci; but when we consider the projective properties of curves, it will be shown that it is possible to project the circular points into a pair of real points, in which case it will usually happen that some of the imaginary foci project into real points. Hence the existence of imaginary foci must not be overlooked, otherwise we should lose sight of various properties connected with the points of intersection of tangents drawn to a curve from a pair of real nodes or cusps.

82. We shall conclude this chapter with two miscellaneous propositions.

To find the equation of the tangents drawn from the point (h, k) to a curve.

Let
$$x\xi + y\eta = 1 \dots\dots\dots\dots\dots\dots\dots(38)$$
be any tangent; and
$$F(\xi, \eta) = 0 \dots\dots\dots\dots\dots\dots\dots(39)$$
the tangential equation of the curve. Since (38) passes through (h, k),
$$h\xi + k\eta = 1,$$
whence by (38),
$$\xi (kx - hy) = k - y,$$
$$\eta (kx - hy) = x - h,$$

whence the equation of the tangents is
$$F \left\{ \frac{k - y}{kx - hy},\ \frac{x - h}{kx - hy} \right\} = 0.$$

83. *A straight line is drawn through a fixed point O; to find the locus of the points of intersection of the tangents at the points where it cuts the curve.*

Let $U = 0$ be the Cartesian equation of the curve referred to O as origin; and let $V = 0$ be the first polar of any point (h, k). Transform to polar coordinates and eliminate r; then the resulting equation will determine $\tan \theta$, where θ is the vectorial angle of the point of contact of any tangent drawn from (h, k). The degree of this equation is necessarily the same as the class of the curve.

Let (h, k) be the point of intersection of the pair of tangents · at any two points P and Q where a straight line through O cuts the curve; then since $\tan \theta = \tan (n\pi + \theta)$ two of the roots of the equation for $\tan \theta$ must be equal; whence the discriminant of this equation equated to zero is the required locus.

4—2

CHAPTER IV.

PLÜCKER'S EQUATIONS.

84. WE have already seen that a cubic curve cannot have more than one double point or a quartic more than three. We shall now give a series of propositions, due to Plücker, by means of which the number and species of the different singularities of a curve of given degree can be determined.

A curve of the nth degree cannot have more than $\frac{1}{2}(n-1)(n-2)$ double points.

Let there be s double points. We have proved in § 16 that when a curve passes through a double point on another curve, it intersects the latter in two coincident points; hence every double point counts for two amongst the points of intersection of two curves. We have also proved in § 35 that the first polar passes through every double point; hence if the first polar intersect the curve in r ordinary points

$$n(n-1) = 2s + r \dots\dots\dots\dots\dots\dots(1).$$

But a curve of the $(n-1)$th degree can be made to satisfy $\frac{1}{2}(n-1)(n+2)$ conditions; if therefore the curve has its maximum number of double points

$$\tfrac{1}{2}(n-1)(n+2) = s + r \dots\dots\dots\dots\dots(2),$$

whence by subtraction

$$s = \tfrac{1}{2}(n-1)(n-2) \dots\dots\dots\dots\dots(3).$$

Equation (3) gives the maximum number of double points for a curve of the nth degree; but we shall hereafter show that if the curve has other singularities, the value of s may be less than the maximum. When $n = 3$, $s = 1$; and when $n = 4$, $s = 3$, as we have proved in Chapter II.

85. *If a curve has δ nodes, the number of its points of inflexion cannot exceed $3n(n-2) - 6\delta$.*

Since any curve which passes through a node on a curve intersects the latter in two coincident points, it follows that if a curve touch one branch of the original curve at a node, the two curves will intersect one another in three coincident points. Similarly if another branch of the second curve touch the other branch of the original curve at the node, the other branch of the second curve will intersect the original curve in three coincident points. Hence if two curves have a common node and two common nodal tangents, they will intersect in six coincident points.

We have shown in § 46 that every node on a curve is a node on the Hessian, and that the two nodal tangents are common to the curve and its Hessian; hence at a node, the curve and its Hessian intersect in six coincident points. We have also shown in § 42 that a curve and its Hessian intersect in $3n(n-2)$ points, and that the Hessian passes through every point of inflexion; if therefore the curve has δ nodes, the curve and its Hessian cannot intersect in more than $3n(n-2) - 6\delta$ ordinary points, and consequently the number of points of inflexion cannot exceed this number.

86. *If a curve has κ cusps, the number of points of inflexion cannot exceed $3n(n-2) - 8\kappa$.*

A cusp may be regarded as the limiting form of a node when the two nodal tangents coincide; hence if A be the cusp and the line $\beta = 0$ be the cuspidal tangent, it follows from § 41 that the equation of the curve must be of the form

$$\beta^2 \alpha^{n-2} + u_3 \alpha^{n-3} + \ldots u_n = 0.$$

By forming the Hessian, it can be shown that the highest power of α is the $(3n-9)$th, and that its coefficient is

$$-2(n-1)\beta^2 d^2 u_3/d\gamma^2,$$

from which it follows that a cusp is a triple point on the Hessian, two of the tangents at which coincide with the cuspidal tangent. But since every branch of a curve which passes through a double point on another curve intersects the latter in two coincident points, it follows that if a double and a triple point coincide the two curves will intersect one another in six coincident points. Also if any branch of the one curve touches any branch of the other curve, the two curves will intersect at a seventh point. But in the present case two of the branches at the triple point on the Hessian touch one another and also the two branches of the cusp on the original curve; accordingly at a cusp the curve and its Hessian intersect one another in eight coincident points, and therefore the number of ordinary points of intersection cannot exceed $3n\,(n-2)-8\kappa$.

By combining the last two theorems it follows that :—

If a curve has δ nodes and κ cusps, the number of points of inflexion is $3n\,(n-2)-6\delta-8\kappa$.

87. *If a curve has δ nodes, the degree of the reciprocal polar cannot exceed $n\,(n-1)-2\delta$.*

We have shown in § 24 that the first polar of a curve with respect to any point O intersects the curve in $n\,(n-1)$ points, which are the points of contact of the $n\,(n-1)$ tangents which can be drawn from O to the curve. Hence the class of a curve, and therefore the degree of the reciprocal polar, cannot exceed this number. We have also shown that the first polar passes through every double point; whence if the curve has δ nodes the first polar intersects the curve in $n\,(n-1)-2\delta$ ordinary points. Hence not more than $n\,(n-1)-2\delta$ tangents can be drawn from O to the curve, which is therefore the degree of the reciprocal polar.

88. *If a curve has κ cusps, the degree of the reciprocal polar cannot exceed $n\,(n-1)-3\kappa$.*

We have shown in § 44 that the first polar touches the curve at a cusp, and consequently at a cusp the curve and its first polar intersect at three coincident points. If therefore a curve has κ cusps, the curve and its first polar cannot intersect at more than $n\,(n-1)-3\kappa$ ordinary points, which is therefore the degree of the reciprocal polar.

By combining the last two theorems, it follows that :—

If a curve has δ nodes and κ cusps, the degree of the reciprocal polar and consequently the class of the curve is $n\,(n-1)-2\delta-3\kappa$.

89.   We are now in a position to establish Plücker's equations. We shall denote

the degree of a curve	by	n,
its class*	,,	m,
the number of its nodes	,,	δ,
,, ,, cusps	,,	κ,
,, ,, double tangents	,,	τ,
,, ,, stationary tangents	,,	ι,
the deficiency of the curve	,,	D.

By §§ 88 and 86, it follows that

$$m = n\,(n-1) - 2\delta - 3\kappa \; \ldots\ldots\ldots\ldots\ldots(4),$$

$$\iota = 3n\,(n-2) - 6\delta - 8\kappa \ldots\ldots\ldots\ldots\ldots(5).$$

We have also shown that a node corresponds to a double tangent on the reciprocal polar, and a cusp to a stationary tangent or tangent at a point of inflexion; also the class of the reciprocal polar is equal to the degree of the original curve and *vice versâ*. Whence reciprocating (4) and (5) we obtain

$$n = m\,(m-1) - 2\tau - 3\iota \; \ldots\ldots\ldots\ldots\ldots(6),$$

$$\kappa = 3m\,(m-2) - 6\tau - 8\iota \ldots\ldots\ldots\ldots\ldots(7),$$

also by § 84

$$D = \tfrac{1}{2}(n-1)(n-2) - \delta - \kappa\ldots\ldots\ldots\ldots\ldots(8).$$

Equations (4) to (8) are Plücker's equations, but only four of them are independent; for if we eliminate δ from (4) and (5) and τ from (6) and (7) the result in both cases is

$$3\,(n-m) = \kappa - \iota\ldots\ldots\ldots\ldots\ldots\ldots(9).$$

---

* Dr Salmon denotes the *degree* of a curve by m and its *class* by n; but since n is usually employed to denote the degree of a curve or of an algebraical expression the notation in the text is preferable.

CHAPTER V.

CUBIC CURVES.

90. THE general equation of a cubic curve contains nine independent constants, that is one less than the number of terms in a ternary cubic; hence a cubic curve may be made to satisfy nine independent conditions. It also follows from § 24 that not more than *six* tangents can be drawn from any external point to the cubic; nor more than *four* from a point on the curve; nor more than *three* from a point of inflexion. Also since a straight line cannot intersect a cubic in more than three points, a cubic cannot have more than one double point unless it breaks up into a conic and a straight line or into three straight lines. Moreover every tangent cuts the cubic at one other point; and since the asymptotes are tangents at infinity, every asymptote cuts the curve at one other point, which may be at a finite or infinite distance from the origin. Also by § 40 a cubic has three asymptotes, one of which must be real.

Cubic curves are divided into the following three species, viz.:

(i) *Anautotomic Cubics*, which have no double point; (ii) *Nodal Cubics*, in which the double point is a crunode or an acnode; (iii) *Cuspidal Cubics*, in which the double point is a cusp. Since $n = 3$, Plücker's numbers for the three species are found by successively putting in equations (4) to (8) of § 89, $\kappa = \delta = 0$; $\kappa = 0$, $\delta = 1$; $\kappa = 1$, $\delta = 0$, which lead to the following table:

n	δ	κ	m	τ	ι	D
3	0	0	6	0	9	1
3	1	0	4	0	3	0
3	0	1	3	0	1	0

91. In § 41 we have discussed several forms of the general equation of a curve of the nth degree in trilinear coordinates, and we shall now consider these special forms when the curve is a cubic.

The general equation may be expressed in the form

$$u_0\alpha^3 + u_1\alpha^2 + u_2\alpha + u_3 = 0 \dots\dots\dots\dots(1),$$

where u_n is a binary quantic in β and γ, or in two other forms in which α, β, γ are interchanged.

The equation of a cubic circumscribing the triangle of reference is

$$\alpha^2 u + \beta^2 v + \gamma^2 w + k\alpha\beta\gamma = 0 \dots\dots\dots\dots(2),$$

where u, v, w are the tangents at A, B and C, and are consequently linear functions of β, γ; γ, α: α, β respectively.

The equation of a cubic having a double point at A is

$$\alpha u_2 + u_3 = 0 \dots\dots\dots\dots(3),$$

also if the cubic pass through the points B and C, u_3 cannot contain β^3 and γ^3; hence the equation of a cubic circumscribing the triangle of reference and having a double point at A is

$$\alpha u_2 + \beta\gamma (\mu\beta + \nu\gamma) = 0 \dots\dots\dots\dots(4).$$

The equation $u_2 = 0$ is the equation of the tangents at the double point; hence the latter will be a node, a cusp or a conjugate point according as the roots of u_2, regarded as a quadratic in β/γ, are real, equal or complex. The line $\mu\beta + \nu\gamma = 0$ is the line drawn from A to the third point where BC cuts the cubic.

If A is a point of inflexion, the tangent at A must meet the cubic in three coincident points. Hence $u_0 = 0$, and u_1 must be a factor of u_2; whence the equation of a cubic having a point of inflexion at A is

$$u_1\alpha^2 + u_1 v_1\alpha + u_3 = 0, \dots\dots\dots\dots(5).$$

92. *If three tangents be drawn to a cubic from a point of inflexion, their points of contact lie on a straight line.*

By (5) the polar conic of A is

$$dF/d\alpha = u_1 (2\alpha + v_1) = 0,$$

and therefore consists of two straight lines, one of which $u_1 = 0$ is the tangent at the point of inflexion A, whilst the other line $2\alpha + v_1 = 0$ passes through the points of contact of the tangents from A. The latter line is called the *Harmonic Polar* of the point of inflexion, and is a line of considerable importance in the theory of cubic curves.

We shall now prove a more general theorem, of which the preceding proposition is a particular case.

93. *If a straight line intersect a cubic in three points D, E, F; the three points D', E', F' in which the tangents at D, E, F intersect the cubic lie on a straight line.*

We shall first prove that every cubic can be expressed in the form
$$uvw + ku'v'w' = 0 \dots\dots\dots\dots\dots\dots(6),$$
where u, v, w and u', v', w' are linear functions of (α, β, γ) and therefore represent three straight lines.

The general equation of a cubic which passes through the vertices of the triangle of reference is
$$\alpha^2 u_1 + \alpha u_2 + \beta\gamma (m\beta + n\gamma) = 0.$$

Add and subtract $l\alpha\beta\gamma$ and the equation becomes
$$\alpha (\alpha u_1 + u_2 - l\beta\gamma) + \beta\gamma (l\alpha + m\beta + n\gamma) = 0;$$

the second term is the product of three straight lines, whilst the first term is the product of a conic and a straight line. Now l may have any value we please; if therefore we determine l so that the discriminant of the conic vanishes, the first term will also be the product of three straight lines.

Equation (6) accordingly represents a cubic passing through the nine points of intersection of (u, v, w) and (u', v', w'). If $u' = v'$, (6) becomes
$$uvw + ku'^2 w' = 0 \dots\dots\dots\dots\dots\dots(7),$$

which is the equation of a cubic which touches the straight lines u, v, w at the points where u' intersects them; also the form of (7) shows that the three points in which u, v, w intersect the cubic lie on the line $w' = 0$.

If D, E, F and D', E', F' be the points in which the lines u' and w' respectively intersect the cubic, the points D', E', F' are called the *tangentials* of D, E, F; and the line $D'E'F'$ is called the *satellite* of DEF.

Since the tangents at the points where the harmonic polar cuts a cubic intersect at a point of inflexion, the tangent at a point of inflexion is the satellite of the corresponding harmonic polar.

94. *The three points in which a cubic intersects its asymptotes lie on a straight line.*

We have shown in § 90 that a cubic has three asymptotes; hence putting $u' = I$, in (7), where $I = 0$ is the line at infinity, the equation

$$uvw + kI^2w' = 0 \quad\ldots\ldots\ldots\ldots\ldots\ldots\ldots(8)$$

is the equation of a cubic of which u, v, w are the asymptotes. The form of this equation shows that the asymptotes intersect the cubic in three points which lie on the straight line $w' = 0$.

The straight line which passes through the points of intersection of a cubic and its asymptotes is called *the satellite of the line at infinity.*

95. *The product of the perpendiculars from any point on a cubic on to the asymptotes, is proportional to the perpendicular from the same point on to the satellite of the line at infinity.*

It follows from (8) that the equation of a cubic referred to a triangle whose sides are the asymptotes is

$$\alpha\beta\gamma + I^2(\lambda\alpha + \mu\beta + \nu\gamma) = 0 \quad\ldots\ldots\ldots\ldots\ldots(9),$$

where (λ, μ, ν) is the satellite of the line at infinity. But if p be the perpendicular from any point of the cubic on to the satellite, p is proportional to $\lambda\alpha + \mu\beta + \nu\gamma$; also I is constant, whence (9) becomes

$$\alpha\beta\gamma = kp.$$

Points of Inflexion.

96. *If a cubic has three real points of inflexion, they lie on a straight line.*

If in (6) we put $u' = v' = w'$, the equation

$$uvw + ku'^3 = 0 \quad\ldots\ldots\ldots\ldots\ldots\ldots(10)$$

represents a cubic having a contact of the second order with the lines u, v, w at the points where the line u' intersects them. Hence the three points of inflexion lie on a straight line.

If the sides of the triangle of reference be the tangents at three real points of inflexion, the equation of the cubic is

$$\alpha\beta\gamma + (l\alpha + m\beta + n\gamma)^3 = 0 \dots\dots\dots\dots\dots(11).$$

In (10) one of the three lines u, v, w must be real, but two of them may be imaginary. In fact the equation $uv_2 + ku'^3 = 0$, where v_2 is any ternary quadric whose discriminant vanishes, represents a cubic one of whose real points of inflexion, and two others which may be real or imaginary, lie on the straight line $u' = 0$. We shall now prove that :—

97. *A cubic cannot have more than three real points of inflexion.*

Let the points B and C be two real points of inflexion, then the third real point of inflexion must lie on this line; hence the equation of the cubic must be

$$(\alpha + \nu\gamma)(\alpha + \mu\beta)(\alpha + m\beta + n\gamma) + l\alpha^3 = 0.$$

Let A be a point on the cubic, then since the equation of the curve cannot contain α^3, $l = -1$ and the equation may be written

$$\alpha^2\{(m + \mu)\beta + (n + \nu)\gamma\} + \alpha\{(m\beta + n\gamma)(\mu\beta + \nu\gamma) + \mu\nu\beta\gamma\}$$
$$+ \mu\nu\beta\gamma(m\beta + n\gamma) = 0.$$

In this equation the coefficient of α^2 is the tangent at A, and must be a *real* straight line.

If possible let A be a real point of inflexion; then it follows from (5) that the coefficient of α^2 must be a factor of that of α, the condition for which is that

$$(m + \mu)(n + \nu)(m\nu + n\mu + \mu\nu) = (n + \nu)^2 m\mu + (m + \mu)^2 n\nu.$$

Putting $\mu x = m$, $\nu y = n$, this equation may be reduced to

$$(1 + x)^2 + (1 + y)^2 = (1 + x)(1 + y),$$

which is a quadratic for determining the ratio $(1 + x)/(1 + y)$; but since its roots are complex, it is impossible to assign real values to μ and ν such that the coefficient of α^2 shall be a factor of that of α; hence A cannot be a real point of inflexion.

The theorems of the last two articles show that the six imaginary points of inflexion of an anautotomic cubic form three conjugate pairs, and that a *real* straight line can be drawn through any conjugate pair and one of the real points of inflexion. It may be added that a pair of conjugate imaginary points are such that the equations of the lines joining them to any vertex (say A) of the triangle of reference are $\beta \pm \iota k\gamma = 0$, so that both lines are included in the equation $\beta^2 + k^2\gamma^2 = 0$.

98. *An acnodal cubic has three real points of inflexion, and a crunodal cubic has one real and two imaginary ones.*

We have shown from Plücker's equations that a nodal cubic cannot have more than three points of inflexion. Let A be the node, C the *real* point of inflexion, BC the tangent at C. Then the equation of the cubic is

$$\beta^3 + (l\beta^2 + 2m\beta\gamma + n\gamma^2)\,\alpha = 0 \ldots\ldots\ldots\ldots(12).$$

Let B' be another point of inflexion, and let $B'C'$ the tangent at B' meet AC in C'. Then if $\beta + k\gamma = 0$ and $\lambda\alpha + \mu\beta + \nu\gamma = 0$ be the equations of AB' and $B'C'$, the equation of the cubic must be

$$(\beta + k\gamma)^3 + (l\beta^2 + 2m\beta\gamma + n\gamma^2)(\lambda\alpha + \mu\beta + \nu\gamma) = 0 \ldots(13).$$

In order that (12) and (13) should represent the same curve we must have

$$k^3 + n\nu = 0,$$
$$3k^2 + n\mu + 2m\nu = 0,$$
$$3k + l\nu + 2m\mu = 0.$$

Eliminating μ and ν, we obtain

$$k\left\{(4m^2 - ln)\,k^2 - 6mnk + 3n^2\right\} = 0.$$

The solution $k = 0$ shows that C is a real point of inflexion, whilst the quadratic factor gives the values of k for the lines joining A to the other two points of inflexion. The condition that these two lines should be real is that $ln > m^2$, and consequently the nodal tangents are imaginary or real according as the other two points of inflexion are real or imaginary.

It frequently happens that when a cubic is drawn the number of real points of inflexion is apparently defective. Whenever this is the case, such singularities exist at infinity which can be found by the methods of §§ 47 to 51.

99. The node of a nodal cubic is the pole of the line joining its three points of inflexion.

The equation of the cubic is

$$\alpha\beta\gamma = (l\alpha + m\beta + n\gamma)^3 = u^3 \text{ (say).}$$

The condition that the cubic should have a double point is obtained by eliminating (α, β, γ) between

$$\beta\gamma = 3lu^2, \quad \gamma\alpha = 3mu^2, \quad \alpha\beta = 3nu^2,$$

from which we deduce

$$l\alpha = m\beta = n\gamma,$$

and the discriminant equated to zero is $27lmn = 1$, which is the condition for a double point.

The polar line of any point (f, g, h) is

$$\alpha\,(gh - 3lu^2) + \beta\,(hf - 3mu^2) + \gamma\,(fg - 3nu^2) = 0,$$

and if this coincides with the line (l, m, n) we must have

$$lf = mg = nh,$$

which shows that (f, g, h) is the node.

The preceding proof holds good when two of the points of inflexion are imaginary, as can at once be seen by writing $\beta + \iota k\gamma$, $\beta - \iota k\gamma$, M and N for β, γ, $m + n$ and $\iota k\,(m - n)$ respectively.

Harmonic Properties.

100. Before commencing to study the harmonic properties of cubics, the following preliminary proposition will be useful.

If a line through the vertex A of the triangle of reference be harmonically divided in P, Q and R; and if the coordinates of these points be denoted by the suffixes 1, 2, 3, then

$$\frac{\alpha_1}{\gamma_1} + \frac{\alpha_3}{\gamma_3} = \frac{2\alpha_2}{\gamma_2}.$$

Let AP cut the base BC in D; let $BAD = \theta$, $BDA = \phi$, then

$$\alpha_1 = (AD - AP) \sin \phi,$$
$$\gamma_1 = AP \sin \theta,$$

and therefore

$$\frac{\alpha_1}{\gamma_1} = \left(\frac{AD}{AP} - 1\right)\frac{\sin \phi}{\sin \theta},$$

whence

$$\frac{\alpha_1}{\gamma_1} + \frac{\alpha_3}{\gamma_3} - \frac{2\alpha_2}{\gamma_2} = AD\left(\frac{1}{AP} + \frac{1}{AR} - \frac{2}{AQ}\right)\frac{\sin \phi}{\sin \theta}$$
$$= 0,$$

since AP is harmonically divided in P, Q and R.

If Q coincides with D, $\alpha_2 = 0$ and the theorem becomes

$$\alpha_1/\gamma_1 + \alpha_3/\gamma_3 = 0,$$

from which it follows that the four lines γ, $\alpha - k\gamma$, α, $\alpha + k\gamma$ form a harmonic pencil. Also if four straight lines form a harmonic pencil, any one of them is called the *harmonic conjugate* of the other three.

101. *Every line through a point of inflexion is divided harmonically by the curve and the harmonic polar.*

Let A be the point of inflexion; and let B and C be two of the points in which the harmonic polar cuts the cubic. Then in (5) we must put

$$u_1 = m\beta + n\gamma, \quad v_1 = 0, \quad u_3 = \beta\gamma\,(\mu\beta + \nu\gamma),$$

and the equation of the cubic becomes

$$(m\beta + n\gamma)\,\alpha^2 + \beta\gamma\,(\mu\beta + \nu\gamma) = 0\dots\dots\dots\dots(14).$$

Let $\beta = k\gamma$ be any line through A which cuts the cubic in P_1 and P_3 and the harmonic polar BC in P_2; substituting the value of β in (14) it becomes

$$(mk + n)\,\alpha^2\gamma + k\gamma^3\,(k\mu + \nu) = 0,$$

whence

$$\alpha_1/\gamma_1 + \alpha_3/\gamma_3 = 0,$$

which shows that

$$\frac{1}{AP_1} + \frac{1}{AP_3} = \frac{2}{AP_2}.$$

102. *Every chord drawn through a point on a cubic is cut harmonically by the curve and the polar conic of that point.*

Let $AP_1P_2P_3$ be the chord cutting the cubic in A, P_1, P_3 and the polar conic of A in P_2. Then the equation of the cubic is

$$\alpha^2 u_1 + \alpha u_2 + u_3 = 0\dots\dots\dots\dots\dots\dots(15),$$

and the polar conic is

$$2\alpha u_1 + u_2 = 0 \quad\ldots\ldots\ldots\ldots\ldots\ldots(16).$$

Let $\beta = k\gamma$ be the equation of AP_1; and let $u_n{}'$ denote what u_n &c. becomes when $\beta = k$, $\gamma = 1$. Then (15) and (16) become

$$\alpha^2 u_1{}' + \alpha\gamma u_2{}' + \gamma^2 u_3{}' = 0,$$

and

$$2\alpha u_1{}' + \gamma u_2{}' = 0,$$

whence

$$\frac{\alpha_1}{\gamma_1} + \frac{\alpha_3}{\gamma_3} = -\frac{u_2{}'}{u_1},$$

$$\frac{\alpha_2}{\gamma_2} = -\frac{u_2{}'}{2u_1},$$

from which it follows that

$$\frac{1}{AP_1} + \frac{1}{AP_3} = \frac{2}{AP_2}.$$

103. *If four tangents be drawn to a cubic from a point A on the curve; and if any line through A intersect the cubic in P and Q, and a pair of opposite chords of contact in D and E; then the line DE is harmonically divided in P and Q.*

Let two of the tangents from A and the corresponding chord of contact be the triangle of reference; then the equation of the cubic is

$$\alpha^2 (m\beta + n\gamma) + \beta\gamma (\lambda\alpha + \mu\beta + \nu\gamma) = 0 \quad\ldots\ldots\ldots(17).$$

The polar conic of A is

$$2\alpha (m\beta + n\gamma) + \lambda\beta\gamma = 0.$$

Multiplying this by $\frac{1}{2}\alpha^2$ and subtracting from (17) it follows that the equation of the chord of contact $B'C'$ of the other two tangents from A is

$$\lambda\alpha + 2\mu\beta + 2\nu\gamma = 0 \quad\ldots\ldots\ldots\ldots\ldots(18).$$

Let $\beta = k\gamma$ be the equation of any line through A; substituting in (17) we find

$$\frac{\gamma_1}{\alpha_1} + \frac{\gamma_2}{\alpha_2} = -\frac{k\lambda}{k\mu + \nu} \quad\ldots\ldots\ldots\ldots\ldots(19).$$

Substituting $k\gamma$ for β in (18) we obtain

$$\frac{\gamma_3}{\alpha_3} = -\frac{k\lambda}{2(k\mu + \nu)},$$

whence

$$\frac{\gamma_1}{\alpha_1} + \frac{\gamma_2}{\alpha_2} = \frac{2\gamma_3}{\alpha_3},$$

from which it can be proved as in § 100 that

$$\frac{1}{DP} + \frac{1}{DQ} = \frac{2}{DE}.$$

104. *If two straight lines be drawn through a point of inflexion to meet a cubic in four points and their extremities be joined directly and transversely, the two points of intersection lie on the harmonic polar.*

Let A be the point of inflexion, and let AB and AC be the two straight lines which meet the cubic in B, D and C, E respectively. Then the equation of the cubic is

$$\alpha\,(m\beta + n\gamma)\,(\lambda\alpha + \mu\beta + \nu\gamma) + \beta\gamma\,(M\beta + N\gamma) = 0 \ldots (20),$$

and the harmonic polar of A is

$$2\lambda\alpha + \mu\beta + \nu\gamma = 0 \ldots\ldots\ldots\ldots\ldots\ldots (21).$$

Let BE, CD intersect in G and BC, DE in H. Putting $\beta = 0$ and $\gamma = 0$ in (20), the equations of BE and CD are

$$\lambda\alpha + \nu\gamma = 0 \text{ and } \lambda\alpha + \mu\beta = 0 \ldots\ldots\ldots\ldots (22),$$

which show that the equation of DE is

$$\lambda\alpha + \mu\beta + \nu\gamma = 0 \ldots\ldots\ldots\ldots\ldots\ldots (23).$$

Equations (21) and (22) show that BE and CD intersect at the point $\lambda\alpha = -\mu\beta = -\nu\gamma$, which by (21) lies on the harmonic polar; whilst (21) and (23) show that DE intersects the harmonic polar at the point where it cuts the line BC.

If AB coincides with AC, the lines BC and DE are the tangents at B and D, whence:—*Tangents at the extremities of any chord drawn through a point of inflexion intersect on the harmonic polar.*

105. *The tangents at any two points of inflexion intersect on the harmonic polar of the point of inflexion which lies on the line joining the other two.*

Let the equation of the cubic be

$$\alpha\beta\gamma + (l\alpha + m\beta + n\gamma)^3 = 0,$$

then if D, E, F be the points in which (l, m, n) cuts BC, CA, AB, then D, E, F are points of inflexion, and BC, CA, AB are the tangents at these points. The coordinates of E are $\beta = 0$, $l\alpha + n\gamma = 0$; whence the polar conic of E is

$$\beta (n\gamma - l\alpha) = 0.$$

The second factor equated to zero is the harmonic polar of E, which obviously passes through B the point of intersection of the tangents at D and F.

106.   *The harmonic polars of three collinear points of inflexion pass through a point.*

By the last article the harmonic polars of the three points D, E, and F are

$$m\beta = n\gamma, \quad n\gamma = l\alpha, \quad l\alpha = m\beta,$$

which obviously meet in a point.

107.   *If a cubic has a double point, each harmonic polar passes through it.*

If A be a point of inflexion, the cubic is given by (5); also if B be a double point, the terms involving β^3 and β^2 must be absent. Whence $v_1 = n\gamma$, $u_3 = \gamma^2 (\mu\beta + \nu\gamma)$; and the harmonic polar of A is $2\alpha + n\gamma = 0$, which obviously passes through B.

Since only one tangent can be drawn from a point of inflexion to a nodal cubic, it follows that the harmonic polar is the line joining the node and the point of contact. When the cubic is cuspidal, the harmonic polar is the cuspidal tangent.

108.   *If two tangents be drawn to a cubic from a point A on the curve, the tangent at the third point where the chord of contact intersects the curve cuts the tangent at A at a point on the curve.*

Let B and C be the points of contact of the tangents from A; let AE be the tangent at A, and DE the tangent at the point D where the chord of contact cuts the curve. Then the equation of the cubic must be of the form

$$\beta\gamma (l\alpha + m\beta + n\gamma) + \alpha^2 (\mu\beta + \nu\gamma) = 0 \ldots\ldots\ldots(24).$$

The form of (24) shows (i) that the line (l, m, n) is the tangent DE at the third point D, where the chord of contact cuts the cubic; (ii) that the line (μ, ν) is the tangent AE at A;

(iii) that the cubic passes through the point of intersection E of (l, m, n) and (μ, ν).

109. *If a chord BDCF, drawn from a point B on a cubic, cut the cubic again in D and C, and the polar conic of B in F; the tangents to the cubic at D and C, and the tangent to the polar conic at F, all pass through the same point.*

Equation (24) shows that (l, m, n) is the tangent at D to the cubic; accordingly if it intersects AC (which is the tangent at C) in G, the equation of BG is $l\alpha + n\gamma = 0$.

The equation of the polar conic of B is

$$\gamma\,(l\alpha + 2m\beta + n\gamma) + \mu\alpha^2 = 0,$$

which shows that the line $(l, 2m, n)$ is the tangent to the polar conic at F. This line obviously intersects AC in G.

110. *If any conic be described through four fixed points on a cubic, the chord joining the two remaining points of intersection of the cubic and the conic will pass through a fixed point on the cubic.*

Let A, B, C, D be the four fixed points on the cubic; let the equations of AD, CD be $\mu\beta + \nu\gamma = 0$ and $\lambda\alpha + \mu\beta = 0$; also let u, v be any linear functions of (α, β, γ). Then the equations of the cubic and the conic may be written

$$\alpha\,(\mu\beta + \nu\gamma)\,u + \gamma\,(\lambda\alpha + \mu\beta)\,v = 0,$$
$$\alpha\,(\mu\beta + \nu\gamma) + k\gamma\,(\lambda\alpha + \mu\beta) = 0,$$

where k is a variable parameter.

The first equation shows that the cubic passes through the point of intersection O of the lines u and v; and dividing the first equation by the second, it follows that the two remaining points of intersection of the cubic and conic lie on the straight line $v = ku$, which obviously passes through O.

The Canonical Form.

111. It is proved in treatises on Algebra* that every ternary cubic whose discriminant does not vanish may be reduced to the canonical form

$$x^3 + y^3 + z^3 + 6lxyz = 0 \quad\ldots\ldots\ldots\ldots\ldots\ldots(25),$$

where (x, y, z) are linear functions of (α, β, γ). We may therefore regard (x, y, z) as the trilinear coordinates of a point referred to

* Elliott's *Algebra of Quantics*, p. 300.

a new triangle of reference, whose sides referred to the original triangle are $x = 0$, $y = 0$, $z = 0$. It therefore follows that every anautotomic cubic curve can be reduced to the above form.

The points where (25) cuts the line $x = 0$ are determined by the equation $y^3 + z^3 = 0$, or

$$(y + z)(y + \omega z)(y + \omega^2 z) = 0,$$

where ω is one of the imaginary cube roots of unity; from which it follows that if l be a variable parameter, all cubics included in (25) cut the three sides of the triangle of reference in the same nine points, three of which are real and six imaginary.

The equation of the tangent at the point $x = 0$, $y = \omega$, $z = -1$ is

$$- 2l\omega x + \omega^2 y + z = 0 \dots\dots\dots\dots\dots(26).$$

Eliminating z between (25) and (26) we obtain $(1 + 8l^3) x^3 = 0$, which shows that if $1 + 8l^3$ is not zero, the line (26) touches the cubic at a point of inflexion. Hence all cubics given by (25), where l is a variable parameter, have the same points of inflexion.

When $1 + 8l^3 = 0$, the discriminant of the cubic vanishes, and the preceding investigation becomes nugatory.

The canonical form being the simplest one to which any ternary cubic, whose discriminant does not vanish, can be reduced is exceedingly useful in a variety of analytical investigations connected with the concomitants of ternary cubic forms; but when discussing the properties of autotomic cubic curves, a special form in which the elements of the triangle of reference have special positions must be employed.

112. *Any cubic, which is described through the nine points of inflexion of another cubic, will have these points for points of inflexion.*

If the cubic U be given by the canonical form (25), the equation of its Hessian H is

$$l^2 (x^3 + y^3 + z^3) - (1 + 2l^3) xyz = 0,$$

whence the Hessian and also the curve $U + \lambda H = 0$ are of the canonical form, where λ is a variable parameter. But this curve represents *any* cubic passing through the nine points of inflexion of U; also by § 111, these points are points of inflexion on $U + \lambda H = 0$.

On the Hessian and the Cayleyan of a Cubic.

113. We have proved in § 38 that if the first polar of any point A has a double point at B, the polar conic of B has a double point at A. In the case of a cubic, the first polar is the polar conic; hence the theorem becomes,—*If the polar conic of A breaks up into two straight lines intersecting at B, the polar conic of B breaks up into two straight lines intersecting at A.* The points A and B obviously lie on the Hessian of the cubic (which is another cubic), and are called by Professor Cayley *conjugate poles*[*]. The envelope of the line joining two conjugate poles was called by Professor Cayley the *Pippian*; but it is now usually known as the *Cayleyan*.

114. *Tangents to the Hessian at two conjugate poles of a cubic intersect on the Hessian.*

Let the conjugate poles A and B be two of the vertices of the triangle of reference; then the polar conics of A and B must be of the form

$$dF/d\alpha = (\alpha + \lambda\gamma)(\alpha + \mu\gamma) = 0,$$
$$dF/d\beta = (m\beta + \lambda\gamma)(\beta + n\gamma) = 0,$$

and therefore the equation of the cubic is

$$\tfrac{1}{3}\alpha^3 + \tfrac{1}{2}(\lambda + \mu)\alpha^2\gamma + \lambda\mu\alpha\gamma^2 + \tfrac{1}{3}m\beta^3 + \tfrac{1}{2}(l + mn)\beta^2\gamma$$
$$+ ln\beta\gamma^2 + \tfrac{1}{3}N\gamma^3 = 0 \dots\dots(27).$$

Now if $\qquad A = d^2F/d\alpha^2, \quad A' = d^2F/d\beta d\gamma$ &c.,

$$\left.\begin{aligned}
A &= 2\alpha + (\lambda + \mu)\gamma \\
B &= 2m\beta + (l + mn)\gamma \\
C &= 2\lambda\mu\alpha + 2ln\beta + 2N\gamma \\
A' &= (l + mn)\beta + 2ln\gamma \\
B' &= (\lambda + \mu)\alpha + 2\lambda\mu\gamma \\
C' &= 0
\end{aligned}\right\} \dots\dots\dots\dots(28),$$

and the equation of the Hessian is

$$H = ABC - AA'^2 - BB'^2 = 0 \dots\dots\dots\dots(29).$$

---

[*] Cayley, "A memoir on curves of the third order." *Phil. Trans.* 1857, p. 415; *Collected Papers*, vol. II. p. 381.

J. J. Walker, *Phil. Trans.* 1888, p. 170; *Proc. Lond. Math. Soc.* vol. XX. p. 382.

The tangents at A and B to the Hessian are the coefficients of α^2 and β^2 in this expression, and are easily seen to be $B = 0$, $A = 0$. These equations obviously satisfy (29), which shows that the tangents at A and B intersect on the Hessian.

For the purpose of simplifying the analysis, we shall take the point C in which the tangents at A and B to the Hessian intersect as the third vertex C of the triangle of reference, in which case the tangents reduce to $\beta = 0$, $\alpha = 0$. This requires that $l + mn = 0$, $\lambda + \mu = 0$, and the cubic becomes

$$\tfrac{1}{3}\alpha^3 - \lambda^2\alpha\gamma^2 + \tfrac{1}{3} m\beta^3 - mn^2\beta\gamma^2 + \tfrac{1}{3} N\gamma^3 = 0 \ \ldots\ldots\ldots(30),$$

whilst the Hessian is

$$\alpha\beta\,(\lambda^2\alpha + mn^2\beta - N\gamma) + (mn^4\alpha + \lambda^4\beta)\,\gamma^2 = 0 \ \ldots\ldots(31).$$

The polar conics of A and B will form a quadrilateral $DEGF$ as shown in the figure; and we shall now prove that:—

115. *The diagonals DG and EF of this quadrilateral intersect at C; and the polar conic of the cubic with respect to C consists of the line AB, and another line passing through the third point K where AB cuts the Hessian.*

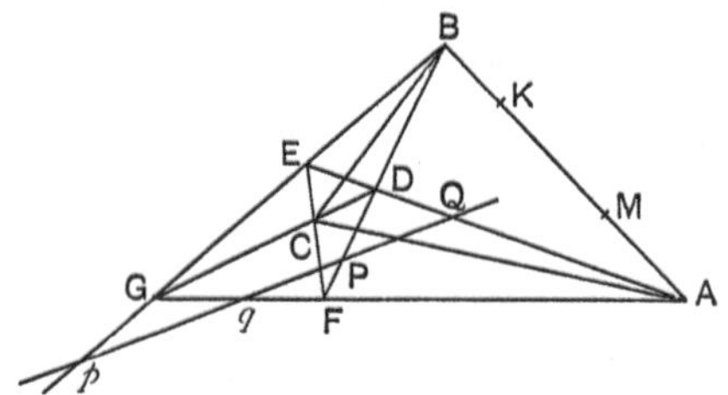

Since the lines BD, BE constitute the polar conic of A, whilst AD, AF constitute that of B, the equation of

$$BD \text{ is } \alpha - \lambda\gamma = 0, \quad BE \text{ is } \alpha + \lambda\gamma = 0,$$
$$AD \text{ is } \beta - n\gamma = 0, \quad AF \text{ is } \beta + n\gamma = 0,$$

from which it follows that the equations of EF and DG are

$$n\alpha + \lambda\beta = 0 \quad \text{and} \quad n\alpha - \lambda\beta = 0,$$

which obviously intersect at C.

To prove the second part, we observe the polar conic of C is

$$- 2\lambda^2\alpha\gamma - 2mn^2\beta\gamma + N\gamma^2 = 0,$$

and therefore consists of the line AB and the line

$$2\lambda^2\alpha + 2mn^2\beta - N\gamma = 0 \quad\ldots\ldots\ldots\ldots\ldots(32).$$

Putting $\gamma = 0$ in (32) the coordinates of K are determined by

$$\lambda^2\alpha + mn^2\beta = 0\ldots\ldots\ldots\ldots\ldots\ldots(33),$$

whence by (31), K is a point on the Hessian.

The harmonic properties of the different lines in the figure are at once evident from § 100.

116. *The polar line with respect to the cubic of any point on the Hessian is the tangent to the Hessian at the conjugate pole.*

The polar line of A is $d^2F/d\alpha^2 = 0$, which by (30) is $\alpha = 0$ or BC, which has already been shown to be the tangent at B to the Hessian.

117. *If M be the point of contact of AB with the Cayleyan, the line AB is harmonically divided at K and M.*

The coordinates of C are $\xi = 0$, $\eta = 0$, $\zeta = b\sin A$. Let $\delta\xi$, $\delta\eta$, $\zeta + \delta\zeta$ be the coordinates of a point C' on the Hessian near C; then the polar conic of C' is

$$\delta\xi\frac{dF}{d\alpha} + \delta\eta\frac{dF}{d\beta} + (\zeta + \delta\zeta)\frac{dF}{d\gamma} = 0 \quad\ldots\ldots\ldots\ldots(34).$$

To find where this intersects AB we must put $\gamma = 0$, and we obtain from (30)

$$dF/d\alpha = \alpha^2, \quad dF/d\beta = m\beta^2, \quad dF/d\gamma = 0,$$

and (34) reduces to

$$\alpha^2\delta\xi + m\beta^2\delta\eta = 0 \quad\ldots\ldots\ldots\ldots\ldots(35).$$

Since C' is a point on the Hessian, $\delta\xi$, $\delta\eta$, $\zeta + \delta\zeta$ satisfy (31); whence writing ξ, η, ζ for α, β, γ in (31), differentiating and putting $\xi = 0$, $\eta = 0$, we get

$$\zeta^2(mn^4\delta\xi + \lambda^4\delta\eta) = 0,$$

and therefore by (35)

$$\lambda^4\alpha^2 = m^2n^4\beta^2,$$

or

$$\lambda^2\alpha \pm mn^2\beta = 0\ldots\ldots\ldots\ldots\ldots\ldots(36).$$

The upper sign furnishes the equation of CK, whilst the lower one gives the equation of CM; whence the lines CA, CM, CK and CB form a harmonic pencil.

When the point A is given, there are in general three conjugate poles corresponding to A; for the tangent at A

intersects the Hessian in one other point C, and from C three other tangents exclusive of the tangent at A can be drawn to the Hessian. The points of contact B, B_1, B_2 of these three tangents are the three conjugate poles; also since the lines AB, AB_1, AB_2 are the only tangents that can be drawn from A to the Cayleyan, this curve is of the third class.

118. *To find the tangential equation of the Cayleyan.*

We have already pointed out that every anautotomic cubic curve may be expressed in the canonical form

$$x^3 + y^3 + z^3 + 6lxyz = 0\dots\dots\dots\dots(37),$$

and that the Hessian is

$$H = -l^2(x^3 + y^3 + z^3) + (1 + 2l^3)\,xyz = 0\dots\dots(38),$$

and is therefore a cubic of the same form as (37).

We have shown in § 115 that the polar conic of C is the line AB and another line through the point K. Let the equations of these lines be

$$\left.\begin{array}{l}\lambda x + \mu y + \nu z = 0\\ \lambda' x + \mu' y + \nu' z = 0\end{array}\right\}\dots\dots\dots\dots(39).$$

Let X, Y, Z be the coordinates of C; then the polar conic of C is

$$X(x^2 + 2lyz) + Y(y^2 + 2lzx) + Z(z^2 + 2lxy) = 0\ \dots(40).$$

In order that (40) may be identical with the product of (39) we must have

$$\lambda\lambda' = kX,\quad \mu\mu' = kY,\quad \nu\nu' = kZ,$$
$$\mu\nu' + \mu'\nu = 2klX,$$
$$\nu\lambda' + \nu'\lambda = 2klY,$$
$$\lambda\mu' + \lambda'\mu = 2klZ,$$

where k is some constant. Eliminating λ', μ', ν' from the last three by means of the first three, we obtain

$$-2l\mu\nu X + \nu^2 Y + \mu^2 Z = 0,$$
$$\nu^2 X - 2l\nu\lambda Y + \lambda^2 Z = 0,$$
$$\mu^2 X + \lambda^2 Y - 2l\lambda\mu Z = 0,$$

whence eliminating X, Y, Z, we get

$$l(\lambda^3 + \mu^3 + \nu^3) + (1 - 4l^3)\lambda\mu\nu = 0\dots\dots\dots(41).$$

This is the tangential equation of the Cayleyan, and its form shows that the curve is of the third class.

If we had eliminated λ, μ, ν and k we should have found that λ', μ', ν' satisfy (41); hence we obtain the theorem :—

The two straight lines which constitute the polar conic of the cubic with respect to any point on the Hessian are tangents to the Cayleyan.

119. From the preceding theorem it appears that the four straight lines AD, AF, BD, BE each touch the Cayleyan, and we shall now prove that :—*The points of contact of these straight lines are collinear.*

Let Q, q be the points of contact of AD, AF; and let $\delta\xi$, $\eta + \delta\eta$, $\delta\zeta$ be the coordinates of a point B' on the Hessian near B. The polar conic of B' is

$$\delta\xi \frac{dF}{d\alpha} + (\eta + \delta\eta)\frac{dF}{d\beta} + \delta\zeta\frac{dF}{d\gamma} = 0.$$

To find where this cuts AD, we must differentiate (30) and put $\beta = n\gamma$, and we obtain

$$dF/d\alpha = \alpha^2 - \lambda^2\gamma^2, \quad dF/d\beta = 0,$$
$$dF/d\gamma = -2\lambda^2\alpha\gamma - 2mn^3\gamma^2 + N\gamma^2.$$

Writing ξ, η, ζ for α, β, γ in (31), differentiating and putting $\xi = \zeta = 0$ we obtain $\delta\xi = 0$; whence the points where the polar conic of B' cuts AD are given by the equation

$$\gamma(2\lambda^2\alpha + 2mn^3\gamma - N\gamma) = 0,$$

and therefore the equation of BQ is

$$2\lambda^2\alpha + 2mn^3\gamma - N\gamma = 0.$$

Putting $\beta = -n\gamma$, it can be shown in the same manner that the equation of Bq is

$$2\lambda^2\alpha - 2mn^3\gamma - N\gamma = 0,$$

whence the points Q, q lie on the straight line

$$2\lambda^2\alpha + 2mn^2\beta - N\gamma = 0 \ \dots\dots\dots\dots\dots\dots(42).$$

By considering the points of intersection with BD, BE of the polar conic of a neighbouring point A', it can be shown that the points P, p lie on (42); whence the four points P, p, Q, q are collinear.

Since equations (32) and (42) are identical, it appears that the four points and also the point K lie on one of the lines which constitutes the polar conic of C.

CHAPTER VI.

SPECIAL CUBICS.

120. In the present chapter we shall consider various special cubics, and shall commence with the discussion of a certain class of circular cubics. It will be shown hereafter that every circular cubic is a degenerate form of a bicircular quartic; hence the theory of circular cubics is best studied as a particular case of these curves. This will be done in Chapter IX.; but the discussion of the circular cubics which are the inverses of conic sections with respect to their vertices deserves separate treatment.

Circular Cubics.

121. *A circular cubic is a cubic which passes through the circular points at infinity.*

From this definition it follows that the trilinear equation of every circular cubic is of the form

$$v_1 S + I v_2 = 0 \quad \dots\dots\dots\dots\dots\dots\dots(1),$$

where S is a circle, I the line at infinity, and v_n is a ternary quantic in α, β, γ. Also since the line v_1 intersects the cubic in two points at a finite distance from the origin and one point at infinity, this line is parallel to an asymptote.

122. *To find the equation of a circular cubic in Cartesian coordinates.*

Since I is a constant, (1) may be written in the form

$$(v_1 + v_0)(r^2 + w_1 + w_0) + V = 0,$$

where V is the general equation of a conic in Cartesian coordinates and v_n, w_n are binary quantics in x and y. This equation is equivalent to

$$v_1 r^2 + u_2 + u_1 + u_0 = 0 \ \dots\dots\dots\dots\dots\dots(2),$$

where u_n is also a binary quantic in x and y. Equation (2) is the general equation of a circular cubic in Cartesian coordinates.

123. *To find the equation of a circular cubic which has a pair of imaginary points of inflexion at the circular points.*

If u, v, w be any three straight lines, the equation

$$u (v^2 + w^2) + I^3 = 0$$

represents a cubic having one real and two imaginary points of inflexion on the line at infinity, and the tangents at the two latter points are $v \pm \iota w = 0$. Let the origin of a system of Cartesian coordinates be the point of intersection of these two tangents, then, if the two imaginary points are the circular points, $v = x$, $w = y$, and the equation of the curve becomes

$$u (x^2 + y^2) + I^3 = 0,$$

or

$$(x^2 + y^2) (px + qy + r) + c^3 = 0,$$

where p, q, r and c are constants. The line $px + qy + r = 0$ touches the curve at the real point of inflexion, which is at infinity; also there will be a node on the axis of x if the discriminant of $x^2 (px + r) + c^3 = 0$ vanishes, which requires that $27 p^2 c^3 + 4r^2 = 0$.

124. *The inverse of a conic with respect to a point on the curve is a circular cubic, whose asymptote is parallel to the tangent to the conic at the centre of inversion.*

The equation of a conic referred to a point on the curve is $u_2 + u_1 = 0$, the inverse of which is $r^2 u_1 + k^2 u_2 = 0$, which is a circular cubic. The origin is obviously a double point, which will be an acnode, a cusp or a crunode, according as the conic is an ellipse, a parabola or a hyperbola; also the line u_1, which is the tangent to the conic at the origin, is parallel to the asymptote of the cubic.

125. We shall now consider the circular cubics which are obtained by inverting a conic with respect to its vertex.

Let the equation of the conic be $x^2/A^2 + y^2/B^2 = 2x/A$; then inverting with respect to a circle of radius k and putting $a = \frac{1}{2}k^2/A$, $b = \frac{1}{2}k^2A/B^2$, the equation of the curve becomes

$$x(x^2 + y^2) = ax^2 + by^2 \quad\dots\dots\dots\dots\dots(3).$$

When a and b are both positive, the curve is the inverse of an ellipse ; when $a = 0$ the curve is the inverse of a parabola and is called a *cissoid*; when b is negative the curve is the inverse of a hyperbola; and when $a = -b$, the curve is the inverse of a rectangular hyperbola and is called the *logocyclic* curve. The latter curve has been discussed by Dr Booth in connection with the geometrical origin of logarithms.

The cubic obviously cuts the axis of x at the origin O, which is a double point, and also at the point A, where $OA = a$, which is called the vertex; and the line $x = b$ is the only real asymptote. The lines $y = \pm (a/3b)^{\frac{1}{2}} x$ cut the curve in two points of inflexion, which are real or imaginary according as the curve is the inverse of an ellipse or a hyperbola. The remaining point of inflexion, which is necessarily real, is at infinity.

The different forms of the curve, according as the conic is an ellipse, a hyperbola or a parabola, are shown in the accompanying figures.

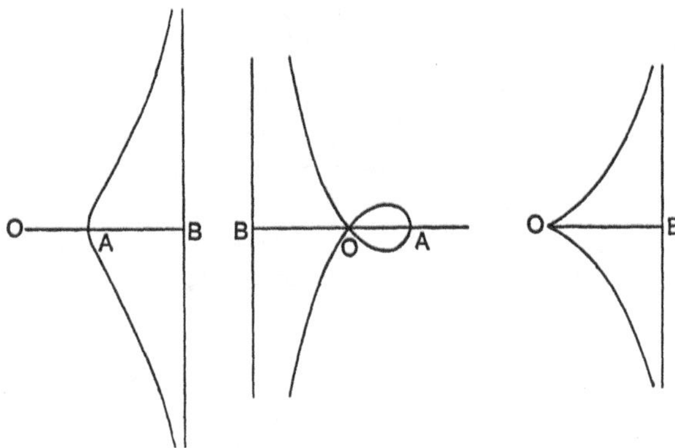

126. *If from the vertex A a straight line is drawn cutting the curve in P, P', then $AP \cdot AP' = AO^2$; and the locus of Q the middle point of PP' is the circular cubic*

$$2x(x^2 + y^2) = (b - a)y^2 - 2ax^2.$$

Transfer the origin to the vertex and then change to polar coordinates and we shall obtain

$$r^2 - r\{(b - a)\sin^2\theta - 2a\cos^2\theta\}\sec\theta + a^2 = 0\dots\dots(4),$$

whence $$AP \cdot AP' = AO^2,$$

and $$2AQ = AP + AP' = \{(b - a) \sin^2 \theta - 2a \cos^2 \theta\} \sec \theta.$$

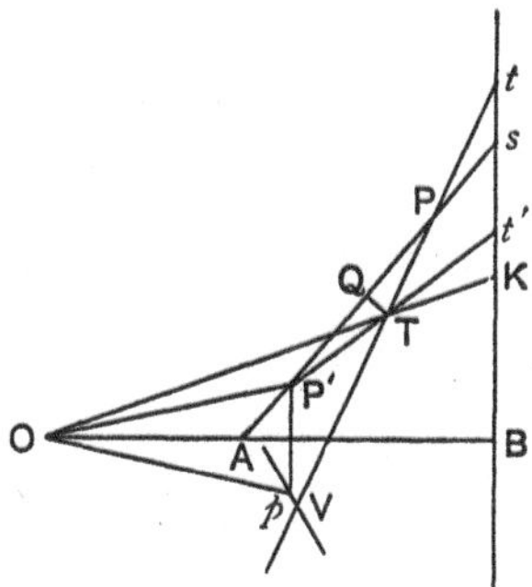

When the cubic is the inverse of a hyperbola, the loop is the inverse of the two branches; but when the cubic is the inverse of an ellipse it follows that if the tangent from A touches the curve at R, $AR = AO$ and the portion between A and R is the inverse of the portion beyond R. Also the locus of Q is a circular cubic of the same species, having a crunode or an acnode according as the signs of $b - a$ and $2a$ are the same or different.

127. *If the tangents at P and P' intersect at T, the locus of T is the cuspidal cubic*

$$x \{(b - a) x^2 + 2by^2\} = 2b^2 y^2.$$

Let $(h,\ k)$ be the coordinates of T referred to O as origin. Transfer the origin to A and let $y = mx$ be the equation of $AP'P$. Then this line must intersect the cubic and the polar conic of T in the points P, P'; if therefore we substitute mx for y we shall obtain two quadratic equations which must be identical.

The cubic leads to the equation

$$x^2 (m^2 + 1) + x \{2a + m^2 (a - b)\} + a^2 = 0,$$

and the polar conic to the equation

$$x^2 \{3h - a + 2km + (h - b) m^2\}$$
$$+ 2x \{2ah - a^2 + k (a - b) m\} + a^2 (h - a) = 0,$$

whence equating coefficients, we get

$$h + km + \tfrac{1}{2} (a - b) m^2 = 0 \quad \dots\dots\dots\dots\dots\dots(5),$$
$$ah + k (a - b) m - \tfrac{1}{2} (a - b) (h - a) m^2 = 0 \quad \dots\dots\dots(6).$$

Eliminating m we get

$$h \left\{ (b-a) h^2 + 2bk^2 \right\} = 2b^2 k^2,$$

which is the equation of the locus.

128. We shall now show that the point T may be found by the following geometrical construction :—

Let AP meet the asymptote in S; bisect BS in K; join OK, and from Q the middle point of PP' draw QT perpendicular to AP meeting OK in T. Then TP, TP' are the tangents at P and P'.

Let OT meet the asymptote in K. Then from (5) and (6) we obtain

$$m = \frac{h^2}{k(b-h)} = \frac{2bk}{h(b-a)}.$$

But

$$m = \frac{BS}{AB} = \frac{BS}{b-a},$$

and

$$\frac{k}{h} = \frac{BK}{b},$$

whence

$$BS = 2BK.$$

Since the points P and P' are inverse points the angle $TPP' = TP'P$; whence $TP = TP'$; hence if TQ be drawn perpendicular to AP, Q is the middle point of PP', and the construction at once follows.

129. *If the tangents at P and P' meet the asymptote in t and t',*

$$Pt = P't'.$$

We have shown in § 126 that

$$2AQ \cos \theta = (AP + AP') \cos \theta = b - a - (b + a) \cos^2 \theta \; ...(7).$$

Also if

$$\phi = TPP' = TP'P,$$

it can be shown from (4) that

$$\cot \phi = \frac{(b + a) \sin \theta + AQ \tan \theta}{PQ}.$$

Now

$$\frac{Pt}{PS} = \frac{\cos \theta}{\cos (\theta + \phi)},$$

whence

$$Pt = \frac{b - a - AP \cos \theta}{\cos (\theta + \phi)}$$
$$= \frac{\{(b + a) \cos \theta + AP'\} \cos \theta}{\cos (\theta + \phi)},$$

by (7).   But

$$\cos (\theta + \phi) = \sin \phi \{(b + a) \cos \theta + AQ - PQ\} \sin \theta / PQ$$
$$= \{(b + a) \cos \theta + AP'\} \sin \theta \sin \phi / PQ,$$

whence

$$Pt = PQ \cot \theta \operatorname{cosec} \phi.$$

Proceeding in the same way, we shall find the same expression for $P't'$; whence $Pt = P't'$.

This proposition was first proved by Dr Booth for the case of the logocyclic curve.

130.   Since nodal circular cubics are curves of the fourth class, it follows that four tangents can be drawn from any point not on the curve. We shall now obtain the quartic equation which determines the vectorial angles of the points of contact.

The polar conic of any point (h, k) is

$$h (3x^2 + y^2 - 2ax) + 2ky (x - b) - ax^2 - by^2 = 0 \ \ldots\ldots(8).$$

Transform (3) and (8) into polar coordinates, eliminate r, and put $z = \tan \theta$, and we shall obtain

$$b (h - b) z^4 + (3bh - 2ab - ah) z^2 - 2k (b - a) z + a (h - a) = 0 \ldots(9).$$

When the point is on the asymptote, $h = b$ and the quartic reduces to a quadratic; whilst if $a = 0$, so that the cubic becomes a cissoid, (9) reduces to a cubic as ought to be the case, since cuspidal cubics are curves of the third class.

131.   *If the ordinate at P' meet the curve again in p, the tangents at P and p intersect on the curve.*

Let V be any point (h, k) on the curve; VP, Vp the tangents drawn from V to the curve.

Equation (9) gives the values of z or $\tan \theta$; but if (h, k) lies on the curve two of the roots of the quartic must be equal to k/h, whence if z_1, z_2 be the other two roots

$$2k/h + z_1 + z_2 = 0 \ \ldots\ldots\ldots\ldots\ldots\ldots(10),$$

$$\frac{k^2 z_1 z_2}{h^2} = \frac{a (h - a)}{b (h - b)},$$

whence by the equation of the curve

$$z_1 z_2 = - a/b \dots\dots\dots\dots\dots\dots(11).$$

Let $POB = \theta$, $pOB = \theta'$; then $z_1 = \tan\theta$, $z_2 = -\tan\theta'$, whence

$$\left. \begin{aligned} \tan\theta - \tan\theta' &= -2k/h \\ \tan\theta\,\tan\theta' &= a/b \end{aligned} \right\} \quad \dots\dots\dots\dots(12).$$

Accordingly

$$\tan\theta + \tan\theta' = 2\,(bk^2 + ah^2)^{\frac{1}{2}}/hb^{\frac{1}{2}} = 2\lambda \text{ (say)}.$$

Produce the ordinate at p to meet the curve in P', then $P'OB = pOB = \theta'$; and the equation of the two lines OP, OP' is

$$ax^2 - 2b\lambda xy + by^2 = 0 \quad \dots\dots\dots\dots(13).$$

The equation of the curve is

$$x\,(x^2 + y^2) - ax^2 - by^2 = 0 \dots\dots\dots\dots(14).$$

Adding (13) and (14) we get

$$x^2 + y^2 - 2b\lambda y = 0 \quad \dots\dots\dots\dots(15),$$

which is the equation of the circle circumscribing the triangle OPP'.

Multiply (15) by a and subtract from (13) and we get

$$(b - a)\,y + 2b\lambda\,(a - x) = 0.$$

This is the equation of the straight line which passes through P and P', and since it is satisfied by $y = 0$, $x = a$, it passes through the vertex A.

132. *The circle circumscribing the triangle OPp passes through a fixed point on the axis.*

From (12) it follows that the equation of OP, Op is

$$ax^2 - by^2 - 2bkxy/h = 0.$$

Subtracting this from (14) we obtain

$$x^2 + y^2 - 2ax + 2bky/h = 0,$$

which is the equation of the circle which passes through OPp. This obviously passes through the point $x = 2a$, $y = 0$.

The Trisectrix of Maclaurin.

133. A particular case of the nodal cubic is the trisectrix of Maclaurin, whose equation is

$$x (x^2 + y^2) = \tfrac{1}{2}a (y^2 - 3x^2),$$

which may be constructed as follows.

Let OCO' be a diameter of a circle whose centre is C; through D, the middle point of OC, draw a straight line perpendicular to OC; draw OBA cutting this line in B and the circle in A; on AO produced take a point P such that $OP = AB$. Then the locus of P is the required curve.

Let $OC = a$; $AOC = \theta$; then

$$- x = OM = OP \cos \theta = AB \cos \theta,$$

and

$$AB = OA - OB = 2a \cos \theta - \tfrac{1}{2}a \sec \theta,$$

whence

$$- x = \tfrac{1}{2}a (4 \cos^2 \theta - 1),$$
$$- y = \tfrac{1}{2}a (4 \cos^2 \theta - 1) \tan \theta,$$

whence eliminating θ, we obtain

$$x (x^2 + y^2) = \tfrac{1}{2}a (y^2 - 3x^2).$$

By means of § 123 it can be shown that the circular points are points of inflexion, that the third point of inflexion (which must be real) is also at infinity, and that the line $x = \tfrac{1}{2}a$ is the inflexional tangent.

The Logocyclic Curve.

134. The logocyclic curve is the inverse of a rectangular hyperbola with respect to a vertex.

Putting $a = -b$ in (3) of § 125, the equation of the curve may be written

$$x (x^2 + y^2) + b (x^2 - y^2) = 0 \dots\dots\dots\dots(1),$$

or
$$r \cos \theta + b \cos 2\theta = 0 \dots\dots\dots\dots(2).$$

The point $x = -b$, $y = 0$ is the vertex, and the line $x = b$ is the asymptote. The form of the curve is shown in the figure.

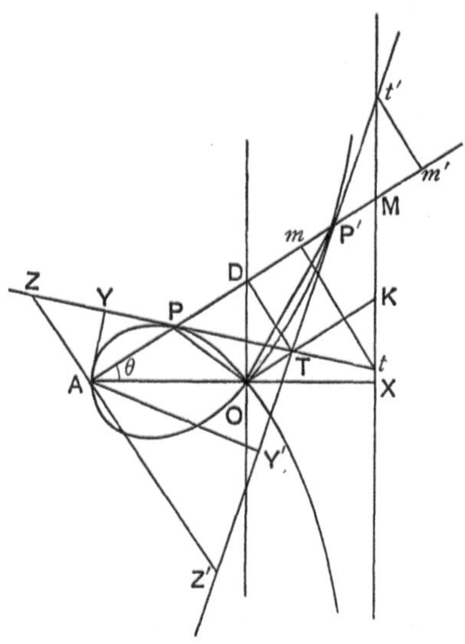

Transfer the origin to the vertex A, and transform to polar coordinates, and the equation of the curve becomes

$$r^2 - 2br \sec \theta + b^2 = 0 \ldots\ldots\ldots\ldots\ldots\ldots(3),$$

whence

$$r = b \, (\sec \theta \pm \tan \theta) \ldots\ldots\ldots\ldots\ldots\ldots(4).$$

Equation (4) enables the curve to be defined in the following manner. Let A be a fixed point, OD a fixed straight line whose distance from A is b, and let AO be perpendicular to OD. Draw any line AD cutting OD in D; and on AD take two points P, P' such that $PD = P'D = OD$. Then the locus of P and P' is the logocyclic curve.

From the construction it follows that

$$AP = b \, (\sec \theta - \tan \theta),$$
$$AP' = b \, (\sec \theta + \tan \theta),$$

whence

$$AP \cdot AP' = b^2 = AO^2.$$

135. *The triangles AOP and $AP'O$ are similar, and the angle POP' is a right angle.*

The first part follows from the relation $AP \cdot AP' = AO^2$, and therefore $AOP = AP'O$. Also since $PD = P'D = OD$,

$$DOP' = DP'O = POA = \tfrac{1}{2}\pi - POD.$$

136. *If ϕ be the angle which the tangent at P makes with AP,*

$$\tan \phi = \cos \theta.$$

Taking the lower sign in (4) we have

$$\tan \phi = - \tan APT = - rd\theta/dr = \cos \theta.$$

From the properties of inverse curves, it follows that ϕ is also the angle which the tangent at P' makes with AP'.

137. *If OK be drawn parallel to AP, and DT be drawn perpendicular to AP meeting OK in T, the lines TP, TP' are the tangents at P and P', and the locus of T is a cissoid.*

Putting $b = - a$ in the result of § 127, it follows that the locus of T is the cissoid $x (x^2 + y^2) = by^2$. Also since $AO = OX$, it follows that the line OK in § 128 is parallel to AP. A direct proof may of course be given.

138. *The locus of the foot of the polar subtangent is a cardioid; whilst that of the polar subnormal is a parabola.*

Let ZAZ' be the polar subtangent; draw AY perpendicular to the tangent at P. Then

$$AZ = AP \tan \phi = b (\sec \theta - \tan \theta) \cos \theta$$

$$= b (1 + \cos ZAO).$$

Also if G, G' be the feet of the polar subnormals

$$AG = AP \cot \phi = b (\sec \theta - \tan \theta) \sec \theta$$

$$= \frac{b}{1 + \sin \theta} = \frac{b}{1 + \cos GAO}.$$

139. *Let the tangents from any point T on the asymptote touch the curve in P and Q; and let the ordinates at these points meet the curve again in P' and Q'; draw $P'O$, $Q'O$ meeting the tangents at Q and P in M and N respectively. Then the angles*

$$PON = QOM = TAO.$$

Equation (9) of § 130 gives the vectorial angles referred to O as origin of the four tangents drawn from any point T. But if T lie on the asymptote $h = b = - a$, and (9) becomes

$$3bz^2 - 2kz - b = 0 \ \dots\dots\dots\dots\dots\dots\dots\dots(5),$$

whence

$$\tan \theta_1 + \tan \theta_2 = 2k/3b,$$
$$\tan \theta_1 \tan \theta_2 = - \tfrac{1}{3},$$

accordingly

$$\tan(\theta_1 + \theta_2) = \tfrac{1}{2}k/b.$$

Now

$$PON = POX - NOX$$
$$= \theta_1 + \theta_2 - \pi,$$

also

$$QOM = QOX - MOX$$
$$= \theta_1 + \theta_2 - \pi,$$

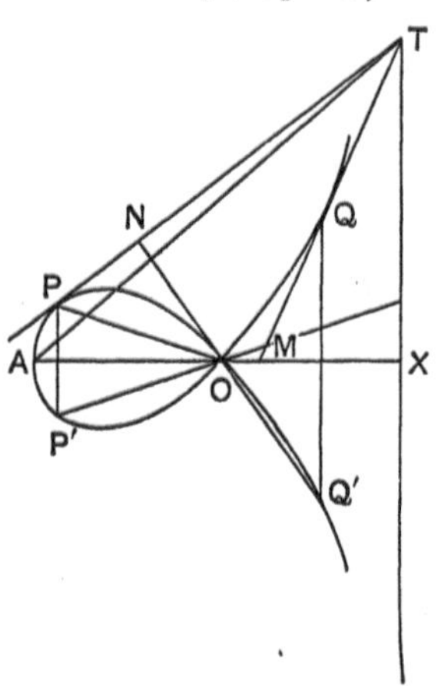

and

$$\tan(\theta_1 + \theta_2 - \pi) = \tan(\theta_1 + \theta_2)$$
$$= \tfrac{1}{2}k/b = TX/AX$$
$$= \tan TAX,$$

whence

$$PON = QOM = TAX.$$

140. *The envelope of the chord of contact PQ is an ellipse.*

Putting $h = b = -a$ in (8) of § 130, the equation of the polar conic is

$$U = 2bx^2 + kxy + b^2 x - bky = 0 \quad \dots\dots\dots\dots(6),$$

and by (5) of § 139, the equation of the two straight lines drawn from the node to the points of contact is

$$V = 3by^2 - 2kxy - bx^2 = 0 \quad \dots\dots\dots\dots(7),$$

whence $U + \lambda V = 0$ is the equation of another conic which passes through the points P, Q and also the origin. The easiest way of determining the condition that this conic should represent two straight lines is to observe that one of them must be of the form

$y = \mu x$; whence substituting and equating coefficients of x we shall find that $\mu = b/k$, $\lambda = k^2/(k^2 - b^2)$, and we obtain

$$(bx - ky)\{(k^2 - 2b^2)x - 3bky + b(k^2 - b^2)\} = 0.$$

The second factor equated to zero is the chord of contact, and its envelope is the ellipse

$$9y^2 + 4(x + b)(2x + b) = 0.$$

Further information on this curve will be found in Booth's *Treatise on some New Geometrical Methods*.

The Cissoid.

141. The cissoid is the inverse of a parabola with respect to its vertex, and its equation is found by putting $a = 0$ in (3) of § 125, and is

$$x(x^2 + y^2) = by^2 \quad\dots\dots\dots\dots\dots\dots\dots(1),$$

or

$$r \cos \theta = b \sin^2 \theta \quad\dots\dots\dots\dots\dots(2).$$

It is also the pedal of the parabola $y^2 + 4bx = 0$ with respect to its vertex.

It is, however, more usual to define the cissoid by the following construction. Let OA be a diameter of a circle, Q any point on its circumference; draw QN perpendicular to OA. Let M be a point on OA such that $OM = AN$, and let MP be drawn perpendicular to OA meeting OQ in P. Then the locus of P is a cissoid.

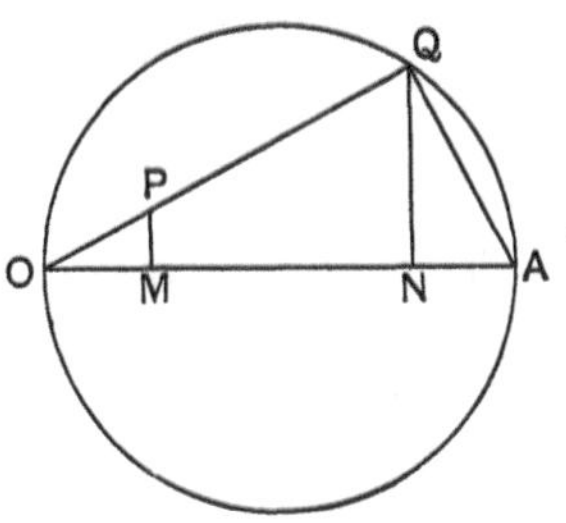

Let $POM = \theta$, $OA = b$; then

$$x = OM = AN = b \sin^2 \theta,$$

$$y^2/(x^2 + y^2) = \sin^2 \theta,$$

whence the locus of P is the curve

$$x(x^2 + y^2) = by^2.$$

The curve has one asymptote, viz. the line $y = b$, also the origin is a cusp; hence the curve is of the third class.

The circle OQA is called the generating circle.

142.   Newton has given the following geometrical construction for drawing a cissoid.

The side CB of a right angle is of constant length $2c$. The side CA passes through a fixed point A, whilst the extremity B moves along a fixed straight line whose distance from A is equal to CB. Then the locus of the middle point P of CB is a cissoid.

Let O be the middle point of AD, then $AO = OD = c$; also let $CAD = \theta$, $x = OM$, $y = PM$. Then

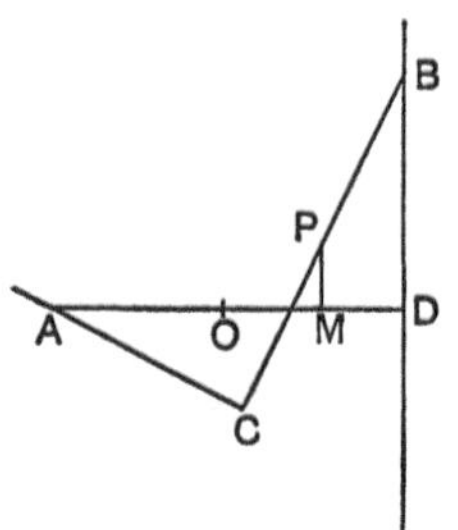

$$c - x = MD = c \sin \theta,$$

$$y \cos \theta + (c + x) \sin \theta = CP = c,$$

whence eliminating θ, we get

$$x (x^2 + y^2) = 2cy^2,$$

and therefore the locus of P is a cissoid.

143.   The cissoid was invented by the Greek geometer Diocles for the purpose of obtaining a geometrical construction for solving the problem of finding two mean proportionals between two straight lines; or, as it is sometimes called, the duplication of the cube. This construction, combined with Newton's method of drawing the curve, enables the problem to be solved by the aid of mechanical appliances.

Let a and b be two straight lines, then it is required to determine x and y such that

$$a/x = x/y = y/b,$$

which requires that $\qquad a^2 b = x^3.$

Let $OA = a$, $OD = b$; join AD meeting the cissoid

$$x (x^2 + y^2) = ay^2$$

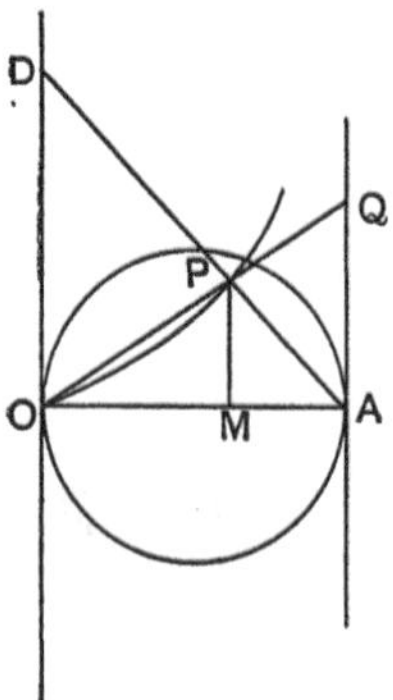

in P. Join OP and produce it to meet the asymptote in Q. Then AQ is the required line.

From the equation of the cissoid

$$\frac{PM^2}{OM^2} = \frac{OM}{AM}.$$

But $\qquad \dfrac{PM}{OM} = \dfrac{AQ}{a},$

and
$$\frac{OM}{AM} = \frac{OP}{PQ} = \frac{b}{AQ},$$

whence
$$\frac{AQ^2}{a^2} = \frac{b}{AQ},$$

or
$$AQ^3 = a^2 b.$$

144. Since a cissoid is a curve of the third class, three tangents can be drawn from any point not on the curve. We shall now explain a geometrical construction by means of which this may be done*.

The polar conic of any point (h, k) is
$$h(3x^2 + y^2) + 2ky(x - b) = by^2 \quad \dots\dots\dots\dots(3).$$

Multiply (1) by $3h$ and (3) by x and subtract and we shall obtain
$$2kx(x - b) = (b + 2h)xy - 3bhy \dots\dots\dots\dots(4).$$

Multiply (3) by $3h$ and (4) by $2k$ and add and we get
$$(9h^2 + 4k^2)x^2 + 3h(h - b)y^2 + 2k(h - b)xy - 4bk^2x = 0.$$

In this write $(x^2 + y^2)/b$ for y^2/x and we get
$$3h(h - b)(x^2 + y^2) + (9h^2 + 4k^2)bx + 2kb(h - b)y - 4b^2k^2 = 0$$
$$\dots\dots\dots(5),$$

which is the equation of the circle passing through the three points of contact of the tangents from (h, k).

A circle and a cissoid intersect in six points, two of which are the circular points at infinity; and we shall now find the fourth point of intersection R.

Transform (5) to polar coordinates, eliminate r by means of (2) and we shall obtain the equation
$$(3h \tan \theta + 2k)\{(h - b) \tan^3 \theta + 3h \tan \theta - 2k\} = 0 \quad \dots(6).$$

Putting $a = 0$ in (9) of § 130, it follows that the second factor gives the vectorial angles of the points of contact of the tangents drawn from (h, k); whence the equation
$$3h \tan \theta + 2k = 0 \quad \dots\dots\dots\dots\dots\dots(7)$$

determines the fourth point R in which the circle cuts the cissoid.

* J. J. Walker, " On tangents to the cissoid," *Proc. Lond. Math. Soc.* Vol. II. p. 161.

The points where (5) cuts the generating circle are found by transforming (5) to polar coordinates and eliminating r by means of the equation $r = b \cos \theta$. This leads to the equations

$$\left. \begin{aligned} 2k \tan \theta_1 + 3h &= 0 \\ 2k \tan \theta_2 - 4h + b &= 0 \end{aligned} \right\} \quad \dots\dots\dots\dots\dots(8).$$

Equations (7) and (8) give the following geometrical construction for drawing three tangents to a cissoid from an external point.

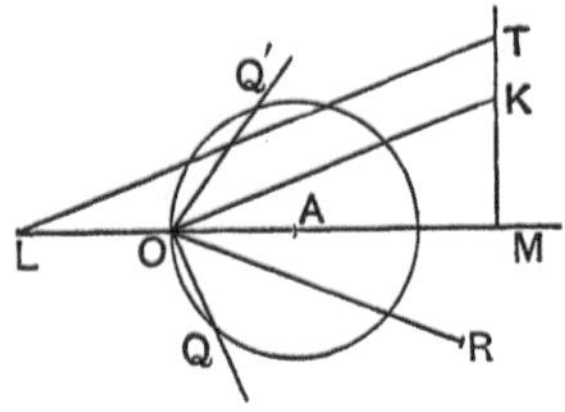

Let A be the centre of the generating circle, OA the cuspidal tangent. From the point T (h, k) draw TM perpendicular to OA, and take K such that $KM = \frac{2}{3}TM$. On the other side of OM draw OR cutting the cissoid in R such that angle $ROM = MOK$. Draw OQ perpendicular to OK meeting the generating circle in Q. Produce AO to L so that $AL = OM$; join LT, and draw OQ' cutting the generating circle in Q', and making with OM an angle $Q'OM = LTM$. Let the circle through $QQ'R$ cut the cissoid in P_1, P_2, P_3; then TP_1, TP_2, TP_3 are the tangents from T.

We have $\qquad \tan ROM = \tan MOK = 2k/3h,$

whence by (7) R is the fourth point of intersection of the cissoid with the circle through the points of contact. Also

$$\tan QOM = \cot MOK = 3h/2k,$$

$$\tan Q'OM = \tan MTL = (4h - b)/2k,$$

whence by (8) Q and Q' are the points where the circle through the points of contact cuts the generating circle.

When the point T is on the curve, the tangent may be drawn by the following simple construction.

Produce the ordinate TM to K such that $KM = 2TM$; join OK and produce it to meet the curve in R, then TR is the required tangent.

Putting $a = 0$ in (9) of § 130, the equation

$$(h - b) \tan^3 \theta + 3h \tan \theta - 2k = 0 \quad \dots\dots\dots\dots(9)$$

determines the vectorial angle of the points of contact of the three tangents drawn from (h, k) to the cissoid. If (h, k) lies on the curve, $k^2(b-h) = h^3$, whence (9) becomes

$$h^3 \tan^3 \theta - 3hk^2 \tan \theta + 2k^3 = 0 \ldots\ldots\ldots\ldots(10),$$

two of the roots of which are equal to k/h (as ought to be the case), whilst the third root is equal to $-2k/h$. This at once gives the foregoing construction.

When T is on the asymptote, $h = b$, and we obtain from (9) $\tan \theta = \tfrac{2}{3}k/h$. Hence if OK meet the curve in P, then P is the required point of contact.

145. *To find the tangential equation of the cissoid.*

The equation of the tangent at (x, y) is

$$X(3x^2 + y^2) + 2Yy(x - b) = by^2 \ \ldots\ldots\ldots\ldots(11),$$

whence
$$\xi = (3x^2 + y^2)/by^2,$$

$$\eta = 2(x - b)/by.$$

Eliminating x and y by means of (1) we obtain

$$27b^2\eta^2 = 4(b\xi - 1)^3 \ \ldots\ldots\ldots\ldots\ldots(12).$$

By § 57, the reciprocal polar is obtained by writing x/k^2, y/k^2 for ξ, η; and is

$$27k^2b^2y^2 = 4(bx - k^2)^3 \ \ldots\ldots\ldots\ldots(13).$$

This curve is the evolute of a parabola.

The pedal of the cissoid with respect to the cusp is obtained by inverting with respect to a circle of radius k, and is

$$27b^2y^2(x^2 + y^2) = 4(bx - x^2 - y^2)^3 \ \ldots\ldots\ldots\ldots(14),$$

and is therefore a sextic curve.

The orthoptic locus is a sextic curve which can be written down by the method of § 68.

Foci.

146. The foci of circular cubics are best studied when the curve is treated as a particular case of a bicircular quartic; we shall therefore only make a few remarks on the subject.

Since nodal circular cubics are of the fourth class, it follows that the curve has one real double focus and four real single ones; also the inverse points of the foci of the conic, whose inverse the curve is, are two of the single foci; and the node is a third focus. On the other hand, cuspidal cubics are of the third class, but in consequence of the cusp replacing the node, the curve has the same number of double and single foci. It will be shown in Chapter VIII. that when a circular cubic has a double point, the latter is a double or a triple focus *composed of the union of two or three single foci*, as the case may be, according as the double point is a node or a cusp; but for the special class of circular cubics considered in the present Chapter a direct proof may be given as follows.

Transform the cubic $x(x^2 + y^2) = ax^2 + by^2$ into trilinear co-ordinates by taking an imaginary triangle of reference, one of whose sides is the line at infinity, whilst the other two sides are the lines joining the double point with the circular points. Then we may write

$$\beta = x + \iota y, \quad \gamma = x - \iota y, \quad I = 1,$$

and the cubic becomes

$$2\beta\gamma\,(\beta + \gamma) = \{a\,(\beta + \gamma)^2 - b\,(\beta - \gamma)^2\}\,I,$$

and therefore the tangents at the circular points (γ, I), (β, I) are

$$2\gamma = I\,(a - b), \quad 2\beta = I\,(a - b),$$

or in Cartesian coordinates

$$2\,(x - \iota y) = a - b, \quad 2\,(x + \iota y) = a - b,$$

which intersect at the point $2x = a - b$, $y = 0$, which determines the double focus.

To obtain the real single foci, we observe that symmetry shows that they must lie on the axis of x; we must therefore find the condition that the line $x - \alpha + \iota y = 0$ should touch the cubic, where $(\alpha, 0)$ are the coordinates of any focus. The points of intersection of this line with the cubic are determined by the equation

$$x^2\,(a - b - 2\alpha) + \alpha\,(2b + \alpha)x - b\alpha^2 = 0,$$

and the line will be a tangent if

$$\alpha^2\,(\alpha^2 - 4b\alpha + 4ab) = 0.$$

In the case of a nodal cubic, the factor $\alpha^2 = 0$ determines two of the single foci, which shows that the node is a double focus formed by the union of two single foci; whilst the other factor determines the two remaining single foci, which are the inverse points of the foci of the conic.

In the case of the cissoid $a = 0$, and the equation becomes

$$\alpha^3 (\alpha - 4b) = 0,$$

which shows that the cusp is a triple focus composed of three single foci, whilst the other single focus is the inverse of the focus of the parabola. The double focus is determined by the equation $x = -\frac{1}{2}b$.

147. *If O be the node of any nodal circular cubic, S and H the two single foci, and P any point on the curve,*

$$l \cdot SP + m \cdot HP = n \cdot OP,$$

where l, m, n are constants. Also if a central conic be inverted with respect to its vertex O, and A be the vertex of the cubic,

$$\frac{SP}{OS} + \frac{HP}{OH} = \frac{OP}{OA}.$$

Let S', H' be the foci of the conic, $2A$ its major axis, P' any point on the conic; then, if unaccented letters denote the inverse points,

$$\frac{SP}{OP} = \frac{S'P'}{OS'} = \frac{S'P' \cdot OS}{k^2},$$

$$\frac{HP}{OP} = \frac{H'P' \cdot OH}{k^2},$$

whence
$$\frac{SP}{OS} + \frac{HP}{OH} = \frac{OP}{k^2}(S'P' + H'P')$$

$$= 2A \cdot OP/k^2,$$

which proves the first part. But when O is the vertex of the conic, $k^2/2A = a = OA$, which proves the second part.

148. The following propositions may be proved by inversion for any nodal circular cubic.

(i) *The circles passing through OSP and OHP cut the curve at equal angles.*

(ii) *If any circle passing through O and a focus cut the cubic in P and Q, the tangent circles at P and Q which pass through O intersect on a fixed circle.*

(iii) *If two circles drawn through O touch the curve at P and Q and also cut one another orthogonally, their other point of intersection lies on a circle.*

We shall conclude this Chapter by discussing a few cubics which do not belong to the foregoing species.

The Semicubical Parabola.

149.   The semicubical parabola is the curve whose equation is

$$ay^2 = x^3 \dots\dots\dots\dots\dots\dots\dots\dots\dots(1),$$

and we shall now prove that this curve is the evolute of a parabola.

Let O be the centre of curvature at any point P of a parabola

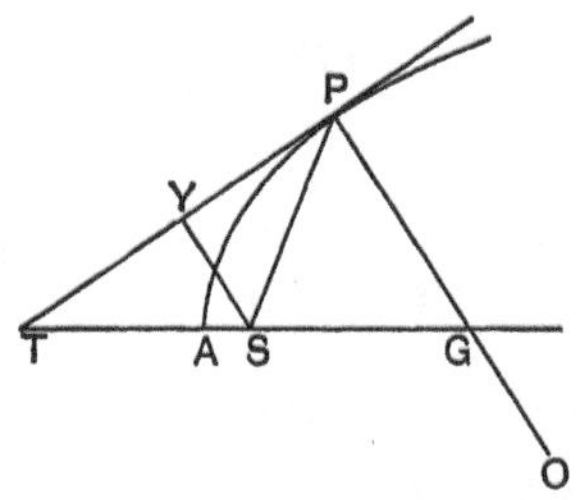

whose latus rectum is $4c$.   Let $SGP = SPG = \psi$; (x, y) the coordinates of O referred to the focus S.   Then by Conics,

$$PO \,.\, SY = 2SP^2,$$

and

$$SP = c \sec^2 \psi, \quad SY = c \sec \psi,$$

$$PO = 2c \sec^3 \psi.$$

Also

$$x = - SP \cos 2\psi + PO \cos \psi$$

$$= c\,(1 + 3 \tan^2 \psi),$$

$$y = - SP \sin 2\psi + PO \sin \psi$$

$$= 2c \tan^3 \psi,$$

whence the locus of O referred to S is

$$27cy^2 = 4\,(x - c)^3 \dots\dots\dots\dots\dots\dots(2).$$

Transfer the origin to the point $x = c$, which is the centre of curvature of the vertex A, and (2) becomes

$$27cy^2 = 4x^3 \ \dots\dots\dots\dots\dots\dots\dots(3),$$

which is of the same form as (1). The form of the curve is shown in the figure.

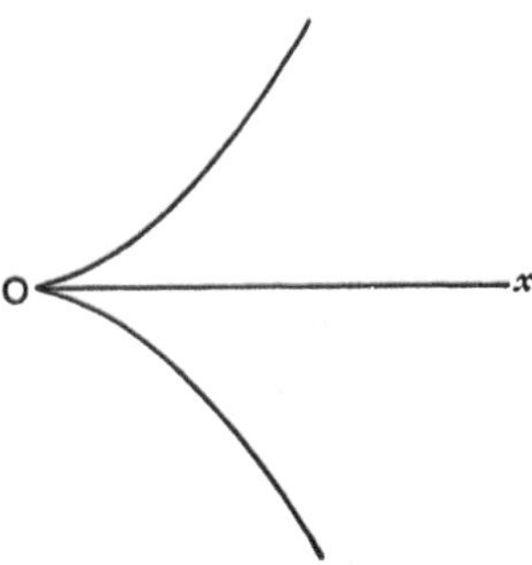

Comparing (2) with (13) of § 145, it follows that the semi-cubical parabola is the reciprocal polar of a cissoid with respect to its cusp, and that the cissoid is the reciprocal polar of a semicubical parabola with respect to the focus of the parabola of which it is the evolute. Also since the cissoid is the inverse of a parabola with respect to its vertex, we may deduce properties of the cissoid from those of the parabola by inversion, and thence deduce properties of the semicubical parabola by reciprocation.

150. *To find the tangential equation referred to the cusp as origin.*

From (1) it follows that the equation of the tangent at (x, y) is

$$3Xx^2 - 2Yay = ay^2,$$

accordingly the tangential equation is

$$4a\xi^3 = 27\eta^2 \quad\dots\dots\dots\dots\dots\dots\dots\dots(4),$$

whence the semicubical parabola is its own reciprocal polar with respect to its cusp. It also follows that the line at infinity is a stationary tangent to the curve; hence the curve has a point of inflexion at infinity, and one real asymptote, viz. the line at infinity.

The first positive pedal is the quartic

$$4ax^3 = 27 (x^2 + y^2) y^2 \quad\dots\dots\dots\dots\dots(5),$$

whilst the orthoptic locus is the parabola

$$y^2 = c (x - c)\dots\dots\dots\dots\dots\dots\dots(6),$$

where $c = 4a/27$, which is a different form of the well known proposition that the locus of the intersection of two perpendicular normals to a parabola is another parabola.

The Cubical Parabola.

151.　The cubical parabola is the curve whose equation is

$$x^3 = a^2 y \quad \dots\dots\dots\dots\dots\dots\dots(1);$$

the origin is therefore a point of inflexion.

The tangential equation of the curve is

$$4a^2 \xi^3 + 27\eta = 0 \quad \dots\dots\dots\dots\dots\dots(2),$$

and consequently the curve is its own reciprocal polar with respect to its point of inflexion.　The curve has a cusp at a point at infinity on the axis of y; and the line at infinity is the cuspidal tangent, and is therefore the real asymptote.

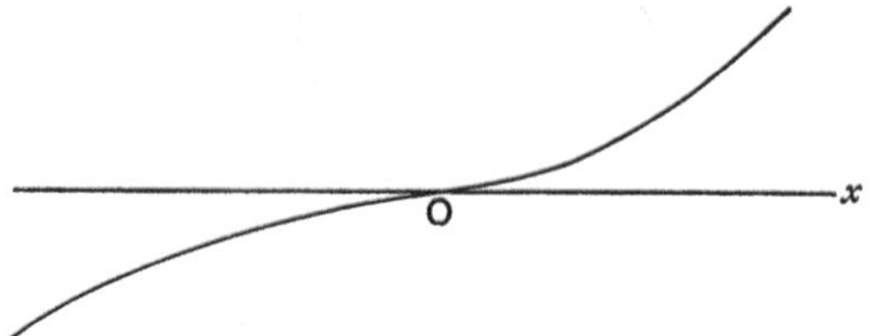

The first positive pedal is the quintic

$$4a^2 x^3 + 27\,(x^2 + y^2)^2\, y = 0 \quad \dots\dots\dots\dots(3),$$

whilst the orthoptic locus is the quartic

$$(xy + c^2)^2 + y^4 = 0,$$

where $c^2 = 4a^2/27$, which breaks up into the axis of x and the rectangular hyperbola $xy + c^2 = 0$.

The Folium of Descartes.

152.　The equation of the folium of Descartes is

$$x^3 + y^3 = 3axy,$$

and if the axes be turned through an angle of 45°, the equation becomes

$$x\,(x^2 + 3y^2) = a\,(x^2 - y^2)\dots\dots\dots\dots\dots(1).$$

The curve may be generated in the following manner:

Let O be a fixed point on a circle whose centre is C, and let CO, CB be two perpendicular diameters.　Draw any line OA

cutting CB in B and the circle in A; and on OA take two points P and Q such that $OP = AB$ and

$$\frac{1}{OQ} + \frac{1}{OB} = \frac{2}{OP} \quad\ldots\ldots\ldots\ldots\ldots\ldots\ldots(2),$$

so that $OQPB$ is a harmonic range; then the locus of Q is the required curve. If a be the radius of the circle, we have

$$x^2 - y^2 = OQ^2 \cos 2\theta,$$

$$x^2 + 3y^2 = OQ^2 (3 - 2\cos^2 \theta),$$

whence
$$\frac{x^2 - y^2}{x^2 + 3y^2} = \frac{\cos 2\theta}{3 - 2\cos^2 \theta} \quad\ldots\ldots\ldots\ldots\ldots\ldots(3).$$

Now
$$OB = a \sec \theta,$$

$$OP = AB = 2a \cos \theta - a \sec \theta \quad\ldots\ldots\ldots\ldots(4),$$

whence by (2)
$$OQ = \frac{a \cos 2\theta}{\cos \theta (3 - 2\cos^2 \theta)},$$

and therefore (3) becomes

$$a (x^2 - y^2) = x (x^2 + 3y^2),$$

which is the locus of Q.

The locus of P is the logocyclic curve, for if (x, y) are the coordinates of P,

$$\frac{x^2 - y^2}{x^2 + y^2} = \cos 2\theta,$$

and by (4)
$$OP \cos \theta = a \cos 2\theta.$$

The form of the curve is almost identical with that of the logocyclic curve. The origin is a crunode, and the line $3x + a = 0$ is the only real asymptote. The curve has one real point of inflexion which is at infinity, and the asymptote is the inflexional tangent. To prove this, interchange x and y in (36) of § 49, and it becomes

$$y (p + qx) (lx + my + n) + Px^3 + Qx^2 + Rx + S = 0.$$

In this, put $m = P = 1$, $Q = -a$, $R = S = l = n = 0$, $p = a$, $q = 3$ and the equation reduces to (1), and the asymptote $3x + a = 0$ is the inflexional tangent at infinity.

The Witch of Agnesi.

153. Let AB and CD be two perpendicular diameters of a circle. Through A draw ANQ, cutting the circle in Q and the diameter CD in N; through Q and N draw QM, NP respectively parallel to CD and AB and intersecting in P. Then the locus of P is a cubic called the witch of Agnesi*.

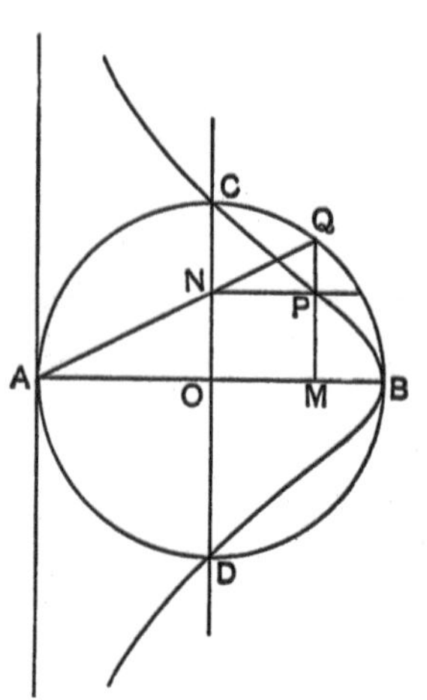

Let A be the origin, $QAM = \theta$; then

$$(y^2 + a^2) \cos^2 \theta = a^2,$$

and

$$x = 2a \cos^2 \theta,$$

whence the equation of the curve is

$$(y^2 + a^2) x = 2a^3 \quad \ldots\ldots\ldots(1).$$

The form of the curve is shown in the figure. It has two real points of inflexion at C and D and a third real point at infinity; also the curve cuts the axis of x at right angles at B, and the axis of y is an asymptote.

The curve has also a conjugate point at infinity, which lies on the axis of x. This result at once follows from (34) of § 48, from which we see that the nodal tangents are determined by $y^2 + a^2 = 0$, and are therefore imaginary.

* Agnesi, *Istituzione analitiche*, Milano 1748. Loria, *Bibliotheca math.* 1897, p. 7.

CHAPTER VII.

CURVES OF THE THIRD CLASS.

154. WE have shown in § 54 that the class of a curve is equal to the degree of its reciprocal polar, also that a node and a cusp respectively correspond to a double tangent and a stationary tangent on the reciprocal polar. Whence curves of the third class are the reciprocal polars of cubics, and may be classified according to the following scheme:

n	δ	κ	m	τ	ι	D
6	0	9	3	0	0	1
4	0	3	3	1	0	0
3	0	1	3	0	1	0

The first species, which are the reciprocal polars of anautotomic cubics, include all sextic curves of the third class. They have nine cusps, and no nodes, double tangents or points of inflexion; also since six of the points of inflexion of an anautotomic cubic must be imaginary, six of the cusps of the sextic must also be imaginary.

The second species, which are the reciprocal polars of nodal cubics, include all quartic curves of the third class. They have three cusps, one double tangent and no nodes or points of inflexion; also since two of the points of inflexion of a crunodal cubic are imaginary, it follows that if the double tangent touches the quartic in two real points, one of the cusps must be real and the two others imaginary. If on the other hand the double tangent touches the quartic in two imaginary points, all three cusps must be real. Since an acnode is a real point, the corresponding double tangent must be a real straight line; but the points of contact, which correspond to the tangents at the acnode, will be imaginary.

The third species consists of cuspidal cubics, which have already been discussed.

155.   We shall now give a few examples of the method by which properties of curves of the third class may be obtained from those of a cubic by reciprocation.

(i)   *A straight line can be drawn through three real points of inflexion of a cubic, or through one real point and two conjugate imaginary ones:* whence,

The cuspidal tangents at three real cusps, or at one real and two conjugate imaginary cusps of a curve of the third class, pass through a point.

Since a quartic curve cannot have more than three cusps, it follows that the cuspidal tangents of a tricuspidal quartic intersect in a point.

(ii)   *If three tangents be drawn to an anautotomic cubic from a point of inflexion, the points of contact lie on a straight line:* whence,

The tangents to the sextic, at the three points where any cuspidal tangent intersects the curve, meet at a point.

This point which is the pole of the harmonic polar will be called the harmonic point of the cuspidal tangent.   In the case of a tricuspidal quartic, the point in question is the point of intersection of the double tangent with the tangent at the point where the corresponding cuspidal tangent cuts the curve; also since a nodal cubic has three harmonic polars which intersect at the node, there are three of such points, which lie on the double tangent of the quartic.

(iii)   *If a straight line intersect a cubic in three points, the three points, in which the tangents at the first three points cut the cubic, lie on a straight line:* whence,

If three tangents be drawn to a curve of the third class from a point, and from the points of contact three other tangents be drawn to the curve, these last three tangents will meet at a point.

(iv)   *If two straight lines be drawn through a point of inflexion to meet a cubic in four points, and their extremities be joined directly and transversely, the two points of intersection lie on the harmonic polar:* whence,

From any two points T, t on a cuspidal tangent of a curve of the third class draw two pairs of tangents TP, TQ and tp, tq to the curve; and let P, p, Q, q be their points of intersection, then PQ and pq pass through the harmonic point.

From (ii) it follows that in the case of a tricuspidal quartic, the lines PQ and pq intersect on the double tangent; which may be easily verified in the case of some simple curve such as the cardioid or the three-cusped hypocycloid.

(v) *If two tangents be drawn to a cubic from a point A on the curve, the tangent at the third point where the chord of contact intersects the cubic meets the tangent at A at a point on the curve:* whence,

Let a straight line touch a curve of the third class at D and intersect it at B and C. Let the tangents at B and C intersect at A, and let the tangent at A touch the curve at E; then DE touches the curve.

156. The foregoing examples sufficiently illustrate the application of the method of reciprocal polars in the case of curves of a higher degree than the second. It will, however, be shown in Chapter XII. that any projective property of a nodal cubic may be deduced from the corresponding property of the logocyclic curve; and therefore instead of reciprocating the properties of this curve, and thereby deriving properties of a special class of tricuspidal quartics, the preferable course is first to generalize by projection, and afterwards to reciprocate. But in the case of properties which are not projective, the method of reciprocation may be employed with advantage in the first instance.

Orthoptic Loci.

157. In § 68 we have explained a general method of finding the orthoptic locus of a curve. We shall now apply this method to examine the orthoptic loci of curves of the third class.

In dealing with this subject, the most convenient classification to make is a fourfold one which is founded upon the position of the origin of reciprocation.

(i) Let the origin *not* lie on the cubic. Then the reciprocal polar consists of all sextic, quartic and cubic curves of the third class which do not touch the line at infinity; and the orthoptic locus is a sextic curve. This may be verified in the case of the cissoid; and it will hereafter be proved that in the case of the cardioid, the orthoptic locus consists of a circle and a limaçon, which together make up a sextic curve.

(ii) Let the origin lie on the curve. Then the reciprocal polar includes all sextic, quartic and cubic curves of the third class which touch the line at infinity; and the orthoptic locus is a quartic curve.

(iii) Let the origin be a node. Then the reciprocal polar includes all quartic curves of the third class to which the line at infinity is a double tangent; and the orthoptic locus is a conic. The three-cusped hypocycloid furnishes an example, for the locus is a circle.

(iv) Let the origin be a cusp. Then the reciprocal polar includes all cubic curves to which the line at infinity is a stationary tangent; and the orthoptic locus is a conic. For example, the orthoptic locus of the evolute of a parabola is a parabola.

CHAPTER VIII.

QUARTIC CURVES.

158. THE general equation of a quartic curve is of the form
$u_4 + u_3 + u_2 + u_1 + u_0 = 0$, where u_n is a binary quantic in x and y,
and therefore contains fourteen independent constants. A quartic
cannot have more than three double points; or it may have two,
one or no double points; also any double point may be a node or
a cusp. It therefore follows from Plücker's formulae, § 89, that
quartic curves may be divided into the following ten species,
which are shown in the accompanying table.

	n	δ	κ	m	τ	ι	D
I.	4	0	0	12	28	24	3
II.	4	1	0	10	16	18	2
III.	4	0	1	9	10	16	2
IV.	4	2	0	8	8	12	1
V.	4	1	1	7	4	10	1
VI.	4	0	2	6	1	8	1
VII.	4	3	0	6	4	6	0
VIII.	4	2	1	5	2	4	0
IX.	4	1	2	4	1	2	0
X.	4	0	3	3	1	0	0

From the preceding table it will be observed first that in the
last four cases the curve is unicursal; secondly, that the tenth
species is the only one in which the quartic is of the third class;
whence a variety of theorems relating to tricuspidal quartics can
be obtained by reciprocating the properties of nodal cubics.
Thirdly, the ninth species is the only one of the fourth class, and
is therefore the only species in which properties of one quartic
can be derived from another by reciprocation.

159. When the equation of a quartic is of the form $u_4 + u_3 = 0$, the origin is a *triple* point, the three tangents at which are given by the equation $u_3 = 0$. Since this is a cubic in y/x, the tangents are (i) all real and distinct, (ii) one real and distinct and two real and coincident, (iii) all real and coincident, (iv) one real and two imaginary. Hence there are four species of triple points; and we shall now show that *every triple point is formed by the simultaneous union of three double points.*

Let A, B, C be three crunodes. When the nodes coincide, the tangents at A and B to the branch AB coalesce into a single tangent. Similarly the tangents at A and C to the branch AC, and those at B and C to the branch BC respectively coalesce into two single tangents. Hence the three pairs of tangents at A, B and C coalesce into three single tangents at the point at which

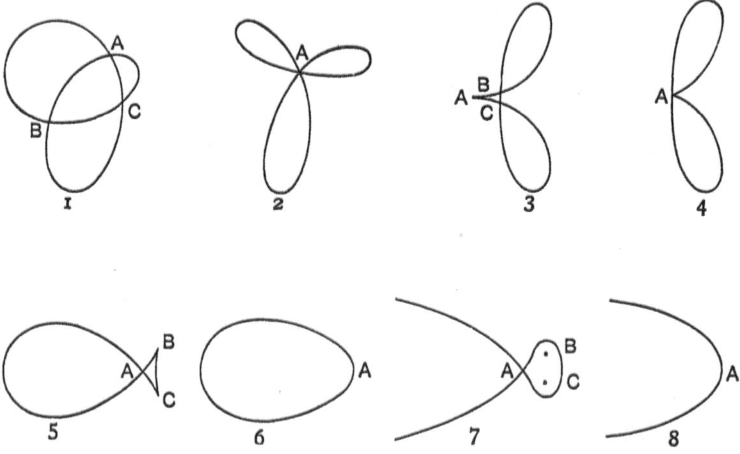

the three nodes ultimately coincide, and therefore this point is a triple point. The forms of the curve before and after union are shown in figures 1 and 2.

The second kind of triple point is composed of two crunodes and a cusp; and the forms of the curve before and after union are shown in figures 3 and 4. The triple point consists of a cusp which lies on the curve.

The third kind of triple point is composed of two cusps and a crunode; and the forms of the curve are shown in figures 5 and 6. The point scarcely differs in appearance from an ordinary point on the curve.

The fourth kind of triple point consists of two conjugate points and a crunode. The forms of the curve are shown in figures 7 and 8, and the point does not differ in appearance from an ordinary point*.

No quartic can have a triple point composed of three cusps; for if such a point existed, the quartic would belong to species X., and therefore its reciprocal polar would be a nodal cubic having three coincident points of inflexion; but on referring to § 98 it will be seen that the equation for k cannot have three roots equal to zero unless n vanishes, in which case the cubic breaks up into three straight lines.

160. Since imaginary singularities occur in pairs, no cubic can have an imaginary node or cusp; but such singularities may occur in all curves of a higher degree than the third. We shall also see that, in addition to the triple point, certain other singularities exist which are formed by the union of two or more simple singularities. We shall therefore require the following additional definitions:

(i) The simple singularities are four in number, viz. the node, the cusp, the double tangent and the stationary tangent.

(ii) A compound singularity is one which is formed by the union of two or more simple singularities. Compound singularities are *real*, *imaginary* or *complex*, according as the simple singularities of which they are composed are all real, all imaginary, or partly one and partly the other.

In the case of an ordinary triple point, the three double points are supposed to move up simultaneously to coincidence; but if two double points *first* move up to coincidence and the third one *afterwards* moves up to coincidence with the first two, we obtain certain singularities which are not triple points. These will now be considered.

Tacnodes.

161. *A tacnode is formed by the union of two nodes.*

In the figure let the two nodes A and B coincide, whilst C remains stationary. The portion ADB, which lies on the side of

* In the case of a quartic, the two conjugate points must lie outside the portion ABC, and must be so situated that no line can be drawn through either of them so as to cut the curve in more than two points.

AB remote from C, thereupon disappears, and the loop $CADB$ touches the branch AB at the point at which A and B coincide. The point A is therefore a tacnode, and the two figures show the forms of the curve just before and just after coincidence. Since the line AB, which ultimately becomes the tangent at A, intersects the curve in two coincident points at A and B, the tangent

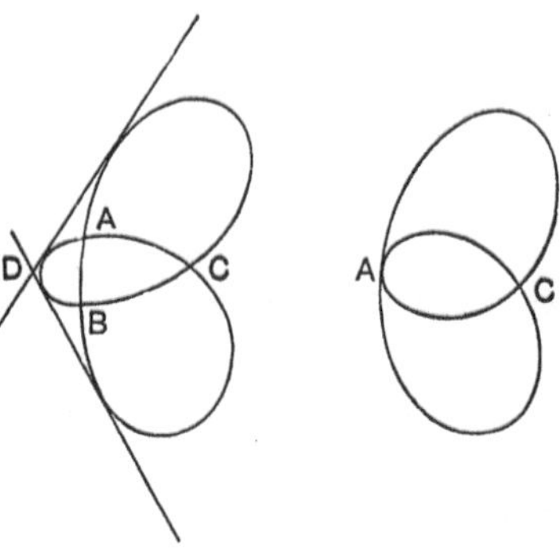

at a tacnode has a contact of the third order with the curve, and therefore cannot intersect the quartic at any other point. Also two double tangents can be drawn, each of which touches the loop CAB; and since they ultimately coincide with the tangent at A, the tacnodal tangent is equivalent to two double tangents. Quartic curves having tacnodes belong to species IV. VII. or VIII., and in each species the number of ordinary double tangents is diminished by 2.

A singular point which is formed by the union of two conjugate points possesses all the properties of a tacnode, but it does not differ in appearance from an ordinary acnode.

The point on the reciprocal curve which corresponds to a tacnode is also a tacnode.

The general equation of a quartic having a node at the origin O is $u_4 + u_3 + u_2 = 0$. If the quartic has another node at a point C whose coordinates are $x = a$, $y = 0$, it follows that when $y = 0$, the quartic must reduce to $x^2 (x - a)^2 = 0$. Also when $x = a$, the resulting equation for y must have one pair of roots equal to zero. Hence the general equation of the quartic must be

$$A x^2 (x - a)^2 + 2Bxy (x - a)(x - b) + y^2 (u_2 + u_1 + u_0) = 0 \ldots \ldots (1),$$

and the equation of the tangents at the origin is

$$A a^2 x^2 + 2Bab xy + u_0 y^2 = 0 \ldots \ldots \ldots \ldots \ldots (2).$$

When the two nodes coincide, $a = 0$, and (1) becomes

$$Ax^4 + 2Bx^2y\,(x - b) + y^2\,(u_2 + u_1 + u_0) = 0 \quad\ldots\ldots\ldots(3),$$

which is the general equation of a quartic having a tacnode at the origin and the axis of x as the tacnodal tangent. The form of this equation shows that the axis of x has a contact of the third order at the tacnode.

The radius of curvature ρ at the origin is the limit of $\tfrac{1}{2}x^2/y$, when x and y vanish; whence putting $x^2 = 2\rho y$ in (3), dividing out by y^2 and then putting $x = y = 0$, we obtain

$$4A\rho^2 - 4Bb\rho + u_0 = 0 \quad\ldots\ldots\ldots\ldots\ldots\ldots(4),$$

whence the two branches lie on the same or on opposite sides of the tacnodal tangent according as u_0/A is positive or negative.

Rhamphoid Cusps.

162. *A rhamphoid cusp is formed by the union of an ordinary cusp and a node.*

The figures show the forms of the curve just before and just after the node and the cusp coincide. It will be observed that the curve possesses one double and one stationary tangent, both of which ultimately coincide with the cuspidal tangent. Hence

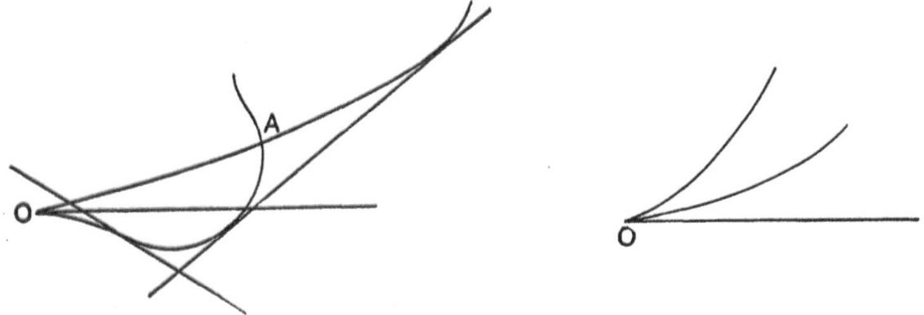

the latter counts once as a double and once as a stationary tangent. Quartic curves having rhamphoid cusps belong to species V. VIII. or IX., and in each species the number of double and stationary tangents is diminished by 1.

The reciprocal polar of a rhamphoid cusp is another rhamphoid cusp.

From (2) it follows that the condition that the origin should be a cusp is that $B^2b^2 = Au_0$; whence by (3) the general equation

of a quartic having a rhamphoid cusp at the origin, and the axis
of x as the cuspidal tangent, is

$$(Ax^2 - Bby)^2 + 2ABx^3y + Ay^2(u_2 + u_1) = 0\ldots\ldots(5),$$

whilst (4) reduces to $(2A\rho - Bb)^2 = 0$, which shows that both
radii of curvature are equal to $\frac{1}{2}Bb/A$.

Oscnodes.

163. *An oscnode is formed by the union of a tacnode and a
node.*

At a tacnode two branches of a curve touch one another; if,
however, the third node C in the figure to § 161 moves up to
coincidence with the tacnode A, the two branches will have a
contact of the second order and will therefore *osculate* one another.
Both branches will therefore have a common circle of curvature at
an oscnode. The forms of the curve before and after union are
shown in the figures. Quartic curves having oscnodes belong to

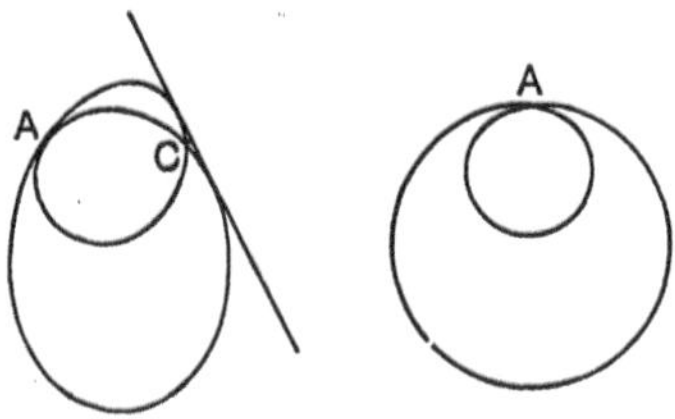

species VII., and have therefore four double tangents; but we
have shown in § 161 that the tangent at a tacnode is equivalent
to two double tangents, and it will be seen from the figure that
the curve has one other double tangent which ultimately coincides
with the oscnodal tangent. Hence the latter is equivalent to
three double tangents, and there is consequently only one ordinary
double tangent. The reciprocal polar of an oscnode is also an
oscnode.

To find the conditions for an oscnode, we observe that (1) is
the equation of a quartic having a node at the origin and at the
point C or $(a, 0)$; we must therefore first find the equation when
the origin is a tacnode, freed from the condition that the axis of x
shall be the tacnodal tangent, and then make the node at C move
up to coincidence with the origin. The following two conditions

must therefore be satisfied, (i) the two tangents at the origin must coincide, (ii) the coincident tangent must have a contact of the third order with the curve.

Equation (2) shows that the first condition requires that $B^2 b^2 = A u_0$, whence (1) may be written

$$\{Ax(x-a) - Bby\}^2 + 2ABx^2 y (x-a) + Ay^2 (u_2 + u_1) = 0 \ldots (6),$$

also by (2) the tangent at the origin is

$$Aax + Bby = 0 \ldots\ldots\ldots\ldots\ldots\ldots (7).$$

Let
$$u_2 = \alpha x^2 + 2\beta xy + \gamma y^2 \left.\right\} \ldots\ldots\ldots\ldots (8).$$
$$u_1 = 2ex + 2fy$$

To find where (7) intersects (6), substitute the value of y from (7), and it will be found that the resulting equation will reduce to $x^4 = 0$, provided

$$B^2 b + Ae - A^2 af/Bb = 0 \ldots\ldots\ldots\ldots (9).$$

Putting $a = 0$ in (6) and (9) and substituting the value of e from (9), (6) becomes

$$(Ax^2 - Bby + Bxy)^2 + y^2 (2Afy + Au_2 - B^2 x^2) = 0 \ldots\ldots (10),$$

which is the general equation of a quartic having an oscnode at the origin and the axis of x as the oscnodal tangent.

Tacnode Cusps.

164. *A tacnode cusp is formed by the union of a tacnode and a cusp.*

The figures show the forms of the curve just before and after coincidence. The curve belongs to species VIII., which has two double tangents; and since the tangent at a tacnode counts twice,

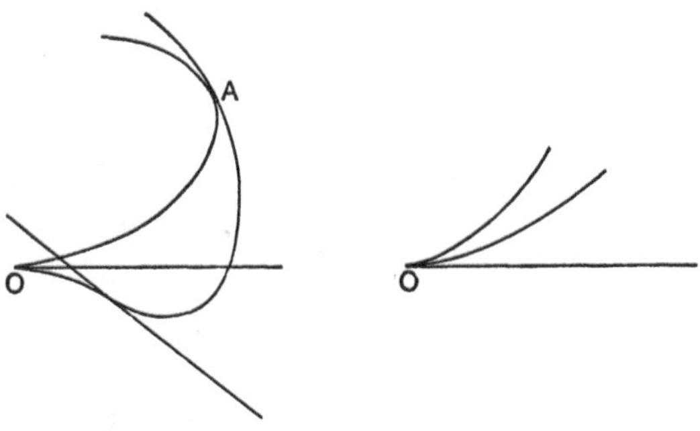

the curve cannot have any ordinary double tangent. The curve
has four stationary tangents, one of which ultimately coincides
with the cuspidal tangent, and consequently there are only three
ordinary stationary tangents.

The reciprocal polar of a tacnode cusp is also a tacnode cusp.

Equation (6) combined with (9) is the equation of a quartic
having a tacnode at the origin, and a double point C at $x = a$,
$y = 0$. To obtain the condition that the origin should be a tacnode
cusp, we must first find the condition that C should be a cusp.
To do this, transfer the origin to C, and pick out the terms of
lowest dimensions, and it will be found that if $x = x' + a$, the
tangents at C are given by the equation

$$A^2 a^2 x'^2 + 2ABa(a - b) x'y + y^2 (B^2 b^2 + A \alpha a^2 + 2A ea) = 0.$$

The condition that the point C should be a cusp is that

$$(B^2 - A\alpha) a = 2 (B^2 b + Ae) \quad\ldots\ldots\ldots\ldots\ldots(11),$$

which by (9) gives

$$B^2 - A\alpha = 2A^2 f / Bb \ldots\ldots\ldots\ldots\ldots\ldots(12).$$

This equation determines the value of f. Substituting in (10),
and changing the constants, it will be found that the resulting
equation may be arranged in the form

$$\{A x^2 - Bby + Bxy + \tfrac{1}{2} (\alpha - B^2/A) y^2\}^2 = Exy^3 + Fy^4 \ldots\ldots(13),$$

which is the general equation of a quartic having a tacnode cusp
at the origin, and the axis of x as the cuspidal tangent.

165. Having explained the nature of the foregoing singularities, we shall now find the trilinear equation of a curve having
one of these singularities at a vertex of the triangle of reference.

Let a quartic have a pair of nodes at A and at a point D on
the line AB; and let the equation of CD be $l\alpha + m\beta = 0$. Then
the equation of the quartic must be of the form

$$\beta^2 (l\alpha + m\beta)^2 + \gamma (\alpha^2 u_1 + \alpha u_2 + u_3) = 0 \ldots\ldots\ldots\ldots(14),$$

where u_n is a binary quantic in β and γ; but since γ^2 must be a
factor when $l\alpha + m\beta = 0$, (14) must be of the form

$$\beta^2 (l\alpha + m\beta)^2 + 2\beta\gamma (l\alpha + m\beta) (\lambda\alpha + \mu\beta) + \gamma^2 (v_0 \alpha^2 + \alpha v_1 + v_2) = 0$$
$$\ldots\ldots\ldots(15).$$

When D coincides with A, $l = 0$, and (15) becomes

$$m^2 \beta^4 + 2m\beta^2 \gamma (\lambda\alpha + \mu\beta) + \gamma^2 (v_0 \alpha^2 + \alpha v_1 + v_2) = 0 \ldots\ldots(16),$$

which is the general equation of a quartic having a *tacnode* at A and the line $\gamma = 0$ as the tacnodal tangent.

The condition that A should be a *rhamphoid cusp* is obtained from (15) by making A a cusp. This requires that $v_0 = \lambda^2$; whence putting $l = 0$, the required equation is

$$(m\beta^2 + \lambda\alpha\gamma)^2 + 2m\mu\beta^3\gamma + \gamma^2 (\alpha v_1 + v_2) = 0 \quad \ldots\ldots(17).$$

By proceeding in the same way as in § 163, it can be shown that the equation of a quartic having an *oscnode* at A is

$$(m\beta^2 + \lambda\alpha\gamma + \mu\beta\gamma)^2 + \gamma^2 (q\alpha\gamma + v_2) = 0 \ldots\ldots\ldots(18),$$

whilst the equation of a quartic having a *tacnode cusp* at A is

$$(m\beta^2 + \lambda\alpha\gamma + \mu\beta\gamma + k\gamma^2)^2 + E\beta\gamma^3 + F\gamma^4 = 0\ldots\ldots\ldots(19).$$

166. *A flecnode is a node, one of the tangents at which is a stationary tangent.*

Since the flecnodal tangent has a contact of the second order with the branch which it touches, and cuts the other branch which passes through the node, every flecnodal tangent has a contact of the third order with the curve.

A biflecnode is a node at which both the tangents are stationary ones.

The lemniscate $(x^2 + y^2)^2 = a^2 (x^2 - y^2)$ has a *real* biflecnode at the origin; and we shall prove hereafter that it has two imaginary biflecnodes at the circular points at infinity.

Flecnodes and biflecnodes may be real, imaginary or complex; but the only complex singularity of this kind is formed by a conjugate point and one or two imaginary stationary tangents.

The reciprocal polar of a flecnode is a double tangent which has a contact of the first order at one point of the reciprocal curve and touches it at a cusp at the other; and the reciprocal polar of a biflecnode is a pair of cusps having a common cuspidal tangent.

Curves of a higher degree than the fourth may have multiple flecnodes, consisting of multiple points, the tangents at which have contacts of higher orders than the second with their respective branches. Thus if a curve of the nth degree has a multiple point of order k, each tangent may have a contact of order $n - k$ or of any lower order with its respective branch.

167. In § 20 a *point of undulation* was defined as *a point where the tangent has a contact of the third order with the curve.* This definition must be understood to mean that the tangent has a contact of the above order at a point which is not one of the preceding singularities. It will be shown in § 180 that the reciprocal singularity is a triple point composed of a node and a pair of cusps.

On curves of the nth degree points exist where the tangent has a contact of any order which is not higher than the $(n-1)$th. Also multiple tangents may exist, which have contacts of orders r, s, t, &c., at different points, where these quantities may have any integral values subject to the condition that

$$r + 1 + s + 1 + t + 1 + \text{&c.}$$

is not greater than the degree of the curve.

Flecnodes and Biflecnodes.

168. We shall now proceed to discuss the properties of flecnodes and biflecnodes of a quartic, but the following preliminary proposition will be useful.

The curve which is the locus of points, whose $(n-r)$th polars break up into a straight line and a curve of degree $r-1$, passes through every point on a curve where the tangent has a contact of the rth order.

The equation of a curve which passes through the vertex A of the triangle of reference is

$$u_1 \alpha^{n-1} + u_2 \alpha^{n-2} + \ldots\ldots u_n = 0 \ldots\ldots\ldots\ldots (1).$$

Now u_1 is the tangent at A, and if this tangent has a contact of the rth order with the curve, u_1 must be a factor of all the u's up to u_r; whence (1) becomes

$$u_1 (v_0 \alpha^{n-1} + v_1 \alpha^{n-2} + \ldots\ldots v_{r-1} \alpha^{n-r}) + \ldots\ldots u_n = 0.$$

The $(n-r)$th polar of A is $d^{n-r}F/d\alpha^{n-r}$, which breaks up into the tangent at A and a curve of degree $r-1$, which proves the proposition. In the case of a quartic, the proposition becomes: *The locus of points, whose polar cubics break up into a conic and a straight line, passes through every point where the tangent has a contact of the third order with the quartic.*

The equation of a quartic having a flecnode at A is

$$\alpha^2 u_1 v_1 + \alpha u_1 v_2 + u_4 = 0 \dots\dots\dots\dots\dots(2),$$

whilst if A is a biflecnode, the equation is

$$\alpha^2 u_2 v_0 + \alpha u_2 v_1 + u_4 = 0 \dots\dots\dots\dots\dots(3).$$

169. *A quartic cannot have more than two flecnodes.*

The equation of a trinodal quartic whose nodes are A, B and C cannot contain any powers of α, β, γ higher than the second, and must therefore be a ternary quadric in $1/\alpha$, $1/\beta$, $1/\gamma$. Hence the required equation is

$$\lambda\beta^2\gamma^2 + \mu\gamma^2\alpha^2 + \nu\alpha^2\beta^2 + \alpha\beta\gamma\,(l\alpha + m\beta + n\gamma) = 0 \ \dots\dots(4).$$

If B and C are flecnodes, the coefficients of β and β^2 must have a common linear factor, and similarly for the coefficients of γ and γ^2; whence the equation of a trinodal quartic having flecnodes at B and C may be written in the form

$$n^2 q\beta^2\gamma^2 + p\gamma^2\alpha^2/(p + q)^2 + l^2 p\alpha^2\beta^2$$
$$+ \alpha\beta\gamma\,\{l\alpha + ln\,(p + q)\,\beta + n\gamma\} = 0 \dots\dots(5).$$

The condition that A should be a flecnode is that the coefficient of α should be a factor of that of α^2. This requires that $p = q$, in which case the quartic becomes a perfect square.

The condition that A should be a cusp is that $p + q = \pm\,2p$. The upper sign must be rejected for the reason stated above; taking the lower sign and changing the constants, (5) may be written

$$\left(\frac{\lambda}{\alpha} + \frac{\mu}{\beta} + \frac{\nu}{\gamma}\right)^2 = \frac{4\lambda}{\alpha}\left(\frac{\lambda}{\alpha} + \frac{\nu}{\gamma}\right) \ \dots\dots\dots\dots\dots(6),$$

which is the equation of a quartic having a cusp at A and a pair of flecnodes at B and C.

170. *If a trinodal quartic has two biflecnodes, the third node must also be a biflecnode. Also two of the biflecnodes must be real and the third one complex; or two must be imaginary and the third real.*

It follows from (3) and (4) that if B and C are biflecnodes the equation of the quartic must be

$$\lambda/\alpha^2 + \mu/\beta^2 + \nu/\gamma^2 = 0 \dots\dots\dots\dots\dots(7),$$

which shows that A must be a biflecnode. In order that the quartic may be real, it is necessary that the sign of one of the constants should be different from those of the other two; whence writing $-\nu$ for ν, it follows that if λ, μ, ν are all positive, the nodal tangents at A and B are real, whilst those at C are imaginary.

To prove the second part, let u, v, w be any real or imaginary straight lines forming a triangle; and consider the quartic

$$\lambda v^2 w^2 + \mu' w^2 u^2 + \nu' u^2 v^2 = 0 \quad\ldots\ldots\ldots\ldots\ldots\ldots(8).$$

Let
$$u = \alpha, \quad v = \beta + \iota k \gamma, \quad w = \beta - \iota k \gamma,$$
$$2\mu' = \mu + \iota \nu, \quad 2\nu' = \mu - \iota \nu,$$

then (8) becomes

$$\lambda\,(\beta^2 + k^2\gamma^2)^2 + \alpha^2\,(\mu\beta^2 - 2k\nu\,\beta\gamma - \mu k^2\gamma^2) = 0 \quad\ldots\ldots(9).$$

Equation (9) represents a quartic having a real biflecnode at A and two imaginary ones at the points where α intersects v and w.

To find what (9) becomes when the imaginary biflecnodes are the circular points at infinity, let A be the origin of a pair of rectangular axes; then since the lines joining A to the circular points are $x \pm \iota y = 0$, we must put

$$\beta = x, \quad \gamma = y, \quad k = 1, \quad \alpha = I$$

in (9), which becomes

$$\lambda\,(x^2 + y^2)^2 + I^2\,\{\mu\,(x^2 - y^2) - 2\nu xy\} = 0,$$

or
$$r^2 = a^2 \cos 2\theta,$$

which is the lemniscate of Bernoulli.

171. We shall now prove that a biflecnode possesses a variety of harmonic properties analogous to those possessed by a point of inflexion on a cubic.

From (3) it follows that the polar cubic of A is

$$u_2\,(2\alpha v_0 + v_1) = 0,$$

and therefore consists of the biflecnodal tangents and the line $2\alpha v_0 + v_1 = 0$. This line, for reasons which will appear in the next section, is called the *harmonic polar* of the biflecnode.

172. *Every line through a biflecnode is divided harmonically by the curve and the harmonic polar.*

Let BC be the harmonic polar; then $v_1 = 0$ and the equation of the quartic becomes

$$\alpha^2 u_2 + u_4 = 0 \quad \dots\dots\dots\dots\dots\dots\dots(10).$$

Let $\beta = k\gamma$ be any line through A; then its points of intersection with (10) are given by

$$\alpha^2 u_2' + \gamma^2 u_4' = 0,$$

where u_2', u_4' are what u_2, u_4 become when $\beta = k$, $\gamma = 1$.   Hence

$$\alpha_1/\gamma_1 + \alpha_2/\gamma_2 = 0,$$

from which it follows from § 100 that the line is divided harmonically by the curve and the harmonic polar.

173. *If two straight lines be drawn from a biflecnode to meet a quartic in four points, and their extremities be joined directly and transversely, the points of intersection will lie on the harmonic polar.*

Let the equation of the quartic be

$$\alpha^2 u_2 - u_4 = 0 \quad \dots\dots\dots\dots\dots\dots\dots(11),$$

where

$$u_2 = (l^2,\ m,\ n^2 \mathbb{X} \beta,\ \gamma)^2,$$
$$u_4 = (\lambda^2,\ \lambda',\ \mu,\ \mu',\ \nu^2 \mathbb{X} \beta,\ \gamma)^4,$$

so that BC is the harmonic polar of the biflecnode A.

Let AB, AC be any two lines through A cutting the quartic in P, Q and p, q respectively; then putting $\gamma = 0$ in (11), the coordinates of P and Q are given by

$$l\alpha = \pm \lambda\beta \dots\dots\dots\dots\dots\dots\dots(12).$$

Putting $\beta = 0$ in (11), the coordinates of p and q are given by

$$n\alpha = \pm \nu\gamma \dots\dots\dots\dots\dots\dots\dots(13).$$

Let the upper signs refer to the points P, p and the lower to the points Q, q; then the equations of Pp and Qq are

$$n(l\alpha - \lambda\beta) - l\nu\gamma = 0,$$
$$n(l\alpha + \lambda\beta) + l\nu\gamma = 0,$$

which obviously intersect on the line BC.   In the same way the equations of Pq and Qp can be shown to be

$$n(l\alpha - \lambda\beta) + l\nu\gamma = 0,$$
$$n(l\alpha + \lambda\beta) - l\nu\gamma = 0,$$

which also intersect on BC.

B. C.   8

If AB and AC coincide, we obtain the theorem that:—*Tangents at the extremities of any chord through a biflecnode intersect on the harmonic polar.*

174.   *The harmonic polar passes through every double point of a quartic.*

In addition to a biflecnode, a quartic may have two other double points; we shall therefore suppose that B is a double point, in which case the terms involving β^3 and β^4 must be absent; whence in (3)

$$u_4 = \gamma^2 v_2$$

and
$$u_2 = \gamma u_1, \ \text{ or } \ v_1 = N\gamma.$$

This value of u_2 is inadmissible, since it would make the quartic break up into a cubic and a straight line; hence $v_1 = N\gamma$ and the harmonic polar is $2\alpha v_0 + N\gamma = 0$, which passes through B.

175.   *Any line through a double point is divided harmonically by the quartic and the polar cubic.*

If A be the double point, the quartic is obtained by putting $u_1 = 0$, $n = 4$ in (1), and the polar cubic is

$$2\alpha u_2 + u_3 = 0 \ \dots\dots\dots\dots\dots\dots\dots(14),$$

which shows that A is a double point on the cubic. If $\beta = k\gamma$ be any chord through A, its points of intersection with the quartic are given by

$$\alpha^2 u_2' + \alpha\gamma u_3' + \gamma^2 u_4' = 0,$$

whence
$$\frac{\alpha_1}{\gamma_1} + \frac{\alpha_2}{\gamma_2} = -\frac{u_3'}{u_2'}.$$

The point of intersection of the chord with the polar cubic is given by

$$\frac{2\alpha_3}{\gamma_3} = -\frac{u_3'}{u_2'},$$

whence
$$\frac{\alpha_1}{\gamma_1} + \frac{\alpha_2}{\gamma_2} = \frac{2\alpha_3}{\gamma_3},$$

which shows that the chord is harmonically divided. This proposition is true when the chord is drawn through any compound singularity which involves a double point.

Points of Inflexion.

176. It appears from § 158 that an anautotomic quartic cannot have more than twenty-four points of inflexion. We shall now prove that the maximum number of *real* points of inflexion is eight.

Let O be a node on a curve; then it follows from §§ 46 and 85 (i) that O is a node on the Hessian, (ii) that the nodal tangents at O are common to the curve and its Hessian, (iii) that the curve and its Hessian intersect in six coincident points at O. Hence each nodal tangent is equivalent to three stationary tangents.

If O is a conjugate point, all six tangents are imaginary; hence a conjugate point reduces the number of imaginary points of inflexion by six.

If O is a real cusp, the curve and its Hessian intersect in eight coincident points at O; hence the cuspidal tangent is equivalent to six imaginary and two real stationary tangents. It therefore follows that a cusp reduces the number of imaginary points of inflexion by six and the number of real ones by two.

If the cusp becomes a crunode, two of the imaginary stationary tangents move away to some other points on the curve, and each nodal tangent is equivalent to one real and two imaginary stationary tangents. Hence a crunode reduces the number of imaginary points of inflexion by four and the number of real ones by two.

If a node or a cusp is imaginary, all the tangents are imaginary; but since imaginary singularities occur in pairs, it follows that a pair of imaginary nodes or cusps reduces the number of imaginary points of inflexion by twelve and sixteen respectively.

177. *To prove that a quartic cannot have more than eight real points of inflexion.*

We have already shown that a crunode reduces the number of *real* points of inflexion by two; hence a *real* biflecnode reduces the number by four. Now if it were possible for a quartic to have *ten* real points of inflexion, the fourteen constants could be determined so that the points A and B should be real biflecnodes, and the point C a real crunode; but we have shown in § 170 that this cannot be done, for if a trinodal quartic has two real biflecnodes the third node must be a complex biflecnode composed of a conjugate point and two imaginary stationary tangents. Hence a quartic cannot have more than eight *real* points of inflexion.

Points of Undulation.

178. We shall commence the consideration of points of undulation by proving the following two theorems.

If the tangent at any point of a curve has a contact of order r, the tangent is equivalent to $r - 1$ stationary tangents.

Let the axis of x have a contact of order r with the curve at the origin ; then the equation of the curve must be of the form

$$y\,(1 + Bx + Cy + \ldots u_{n-1}) + x^{r+1}\,(a + bx + cy + \ldots v_{n-r-1}) = 0,$$

where u_n, v_n are binary quantics in x and y.

A first approximation shows that the form of the curve in the neighbourhood of the origin is $y + ax^{r+1} = 0$; whilst a second approximation gives

$$y + x^{r+1}\,\{a + (b - B)\,x\} = 0,$$

whence

$$-\frac{d^2y}{dx^2} = r\,(r + 1)\,x^{r-1}\,\{a + (b - B)\,x\} + 2\,(r + 1)\,x^r\,(b - B).$$

If there is a point of inflexion at a point Q in the neighbourhood of the origin, the abscissa of Q will be given by the equation $d^2y/dx^2 = 0$, and is therefore

$$x = -\frac{ar}{(r + 2)\,(b - B)}.$$

When Q moves up to coincidence with the origin, $a = 0$, and consequently when $y = 0$, the equation of the curve reduces to

$$x^{r+2}\,(p_0 + p_1 x + \ldots p_{n-r-2}\,x^{n-r-2}) = 0,$$

which shows that the axis of x has a contact of order $r + 1$ at the origin.

The preceding theorem shows that every point of undulation is formed by the union of two points of inflexion; and also, in combination with § 177, shows that a quartic cannot have more than twelve points of undulation, and that not more than four of these points can be real.

179. *Every quartic may be expressed in the form $S^2 = uvwt$, where S is a conic, and u, v, w, t are straight lines.*

The general equation of a quartic may be written in the form of a ternary quadric in U, V, W, where these quantities equated to zero represent three conics. The simplest way of proving this is to recollect that every ternary quadric can be expressed as the sum of three squares by means of a linear transformation. The quartic can accordingly be expressed in the form

$$lU^2 + mV^2 + nW^2 = 0,$$

and if the terms be multiplied out it will be found that the equation contains fourteen independent constants. It also follows that any form of the equation of a conic in trilinear coordinates will represent a quartic if U, V, W be substituted for α, β, γ.

Every quartic may be regarded as the envelope of the conic $\lambda^2 U + 2\lambda V + W = 0$, where λ is a variable parameter; for the envelope is the quartic $V^2 = UW$, which by the last paragraph is one of the forms to which every quartic may be reduced.

The equation $V^2 = UW$ is equivalent to the equation

$$\{\lambda\mu U + (\lambda + \mu) V + W\}^2 = (\lambda^2 U + 2\lambda V + W)(\mu^2 U + 2\mu V + W),$$

where λ and μ are arbitrary constants, as can at once be seen by multiplying out. The left-hand side is the square of a conic, and by determining λ and μ so that the discriminants of the two factors on the right-hand side vanish, the latter may be reduced to four linear factors. Hence any quartic may be reduced to the form $S^2 = uvwt$, where S is a conic and u, v, w, t are four straight lines. This form is due to Plücker, and furnishes a means of determining the double tangents to a quartic.

180. We have shown in the last article that every quartic may be written in the form

$$S^2 + uvwt = 0 \quad \dots\dots\dots\dots\dots\dots\dots(1).$$

The four straight lines u, v, w, t obviously touch the quartic at the eight points where the conic cuts it, and are therefore double tangents to the quartic. If, however, u, v, w, t touch S, the points of contact will be points of undulation on the quartic, and they may be all real, all imaginary, or two real and two imaginary. From this it follows that every tangent at a point of undulation is equivalent to one double tangent; but we have shown in § 178 that it is also equivalent to two stationary tangents, whence the reciprocal singularity is a triple point composed of a node and a pair of cusps.

181.   Let (1) be written in the form

$$\alpha\beta\gamma u + S^2 = 0 \quad \dots\dots\dots\dots\dots\dots(2),$$

where

$$S = l^2\alpha^2 + m^2\beta^2 + n^2\gamma^2 - 2mn\beta\gamma - 2nl\gamma\alpha - 2lm\alpha\beta \dots\dots(3),$$

$$u = \lambda\alpha + \mu\beta + \nu\gamma \quad \dots\dots\dots\dots\dots\dots(4),$$

$$l/\lambda + m/\mu + n/\nu = 0 \quad \dots\dots\dots\dots\dots\dots(5),$$

then the conic S touches the quadrilateral α, β, γ, u at four real points which are real points of undulation on the quartic.

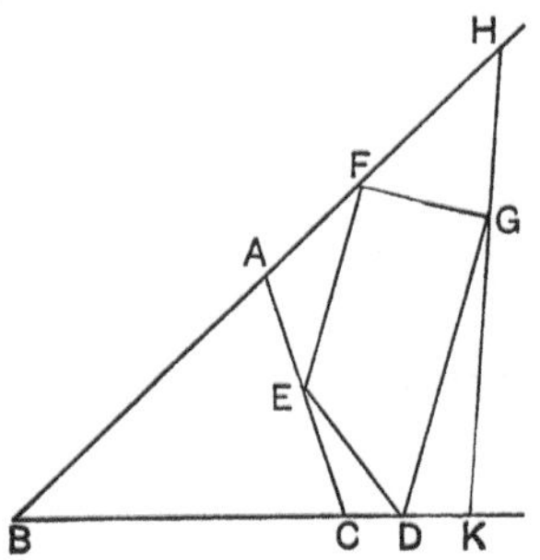

In the figure, ABC is the triangle of reference; HK is the line $u = 0$, and D, E, F, G are the four points of undulation. The coordinates of G are obtained by solving the equations $S = 0$, $u = 0$, and are determined by

$$\lambda^2\alpha/l = \mu^2\beta/m = \nu^2\gamma/n \quad \dots\dots\dots\dots\dots(6),$$

whilst the equation of DG is

$$\lambda\left(\frac{m}{\mu} - \frac{n}{\nu}\right)\alpha + m\beta - n\gamma = 0 \quad \dots\dots\dots\dots(7).$$

We notice that the three straight lines AD, BE, CF meet at the point $l\alpha = m\beta = n\gamma$. Also four triangles can be formed by taking any three of the four straight lines AB, BC, CA, HK, and any of these triangles may be taken as the triangle of reference. We thus obtain the following theorem :—

If a triangle be formed by the tangents at any three real points of undulation, the lines joining the vertices of the triangle with the points of contact of the opposite sides meet at a point.

182. By a rearrangement of terms (2) can be written

$$\beta\gamma\Sigma_1 = EF^4 \dots\dots\dots\dots\dots(8),$$

where $EF = l\alpha - m\beta - n\gamma$, so that $EF = 0$ is the equation of EF, and

$$\Sigma_1 = 8mnEF^2 - 16m^2n^2\beta\gamma - \alpha u \dots\dots\dots(9).$$

The form of (8) shows that the conic Σ_1 has a contact of the third order with the quartic at two points E_1, F_1 which lie on EF; but these points will not necessarily be points of undulation, since the only condition that has been imposed is that the quartic should have four real points of undulation, and if the quartic is to have more than four such points a further relation between the constants is necessary. The required condition is that the discriminant of Σ_1 should vanish, in which case the conic breaks up into two straight lines which touch the quartic at two more points of undulation.

The conic Σ_1, when written out at full length, becomes

$$(8l^2mn - \lambda)\,\alpha^2 + 8mn\,(m^2\beta^2 + n^2\gamma^2)$$
$$- (16lmn^2 + \nu)\,\gamma\alpha - (16lm^2n + \mu)\,\alpha\beta = 0\dots\dots(10),$$

but it will be more convenient to consider the discriminant of the conic

$$(2l^2k - \lambda)\,\alpha^2 + 2k\,(m^2\beta^2 + n^2\gamma^2) - (4mnk - k^2)\,\beta\gamma$$
$$- (4nlk + \nu)\,\gamma\alpha - (4lmk + \mu)\,\alpha\beta = 0\dots\dots(11),$$

which reduces to (10) when $k = 4mn$.

The discriminant of (11), when equated to zero, leads to a quintic equation which contains k as a factor. The quartic factor is resolvable into two quadratic factors which furnish the equations

$$k^2 - 8kmn + \mu\nu/\lambda = 0\dots\dots\dots\dots(12),$$
$$k^2 - 8kmn - k\lambda/2l^2 + \mu\nu/\lambda = 0 \dots\dots\dots(13).$$

Since $k = 4mn$, it follows from the first equation that

$$16m^2n^2 = \mu\nu/\lambda \dots\dots\dots\dots(14),$$

which in combination with (5) reduces (10) to

$$m^2\,(\lambda\alpha + \mu\beta)^2/\nu^2 + n^2\,(\lambda\alpha + \nu\gamma)^2/\mu^2 = 0 \dots\dots(15).$$

If this result were admissible, (15) would represent a pair of imaginary straight lines touching the quartic at a pair of imaginary points of undulation which lie on EF; but since these lines intersect at the real point

$$- \lambda\alpha = \mu\beta = \nu\gamma,$$

which by virtue of (5) lies on EF, this result is impossible ; hence the relation (14) between the constants is inadmissible, and we must therefore consider the second equation (13).

Writing $k = 4mn$, (13) becomes

$$\frac{\mu\nu}{4\lambda mn} = 4mn + \frac{\lambda}{2l^2} \ \dots\dots\dots\dots\dots(16),$$

which by (5) may be expressed in the form

$$\frac{16l^2m^2n^2}{\lambda\mu\nu} = \frac{m^2}{\mu^2} + \frac{n^2}{\nu^2} \ \dots\dots\dots\dots\dots(17).$$

Now (10) may be written in the form

$$\left(l^2 - \frac{\lambda}{8mn}\right)\alpha^2 + m^2\beta^2 + n^2\gamma^2 - 2\left(l + \frac{\nu}{16mn^2}\right)n\gamma\alpha$$

$$- 2\left(l + \frac{\mu}{16m^2n}\right)\alpha\beta = 0\dots\dots\dots(18),$$

also by means of (16), (13) and (5) it can be shown that

$$l^2 - \frac{\lambda}{8mn} = \left(l + \frac{\mu}{16m^2n}\right)^2 + \left(l + \frac{\nu}{16mn^2}\right)^2,$$

accordingly (18) reduces to

$$\left\{\left(l + \frac{\mu}{16m^2n}\right)\alpha - m\beta\right\}^2 + \left\{\left(l + \frac{\nu}{16mn^2}\right)\alpha - n\gamma\right\}^2 = 0 \ \dots(19),$$

which is the equation of a pair of imaginary straight lines touching the quartic at two imaginary points of undulation which lie on EF.

183. The equation of the conic S may be written in either of the forms

$$S = EF^2 - 4mn\beta\gamma,$$

or

$$S = DG^2 + 4mn\lambda\alpha u/\mu\nu,$$

and consequently (2) may be written in the form

$$\alpha u\Sigma_2 + DG^4 = 0,$$

where

$$\Sigma_2 = \left(\frac{m^2}{\mu^2} + \frac{n^2}{\nu^2}\right)\lambda^2\alpha^2 + m^2\beta^2 + n^2\gamma^2 + 2\left(\frac{\mu\nu}{16\lambda mn} - mn\right)\beta\gamma$$

$$+ \frac{2n^2\lambda}{\nu}\gamma\alpha + \frac{2m^2\lambda}{\mu}\alpha\beta = 0 \ \ldots\ldots(20).$$

Equation (20) will break up into two straight lines if the coefficient of $\beta\gamma$ vanishes; but we have shown that this relation between the constants is inadmissible. We shall therefore prove that either of the equivalent equations (16) or (17) resolves (20) into the product of two linear factors.

Equation (16) reduces the coefficient of $\beta\gamma$ to $\frac{1}{4}\lambda/l^2$, from which it can easily be shown that (20) may be expressed in the form

$$\left(\frac{m\lambda}{\mu}\alpha + m\beta + \frac{\lambda}{8l^2m}\gamma\right)^2 + \left(\frac{n\lambda}{\nu}\alpha + n\gamma - \frac{\lambda\nu}{8l^2n\mu}\gamma\right)^2 = 0,$$

which represents a pair of imaginary straight lines touching the quartic at two points of undulation which lie on DG.

184. It thus appears that when the constants are connected together by the relation (17), the quartic has four imaginary points of undulation which lie in pairs on the lines EF and DG respectively. In the same way if

$$\frac{16l^2m^2n^2}{\lambda\mu\nu} = \frac{l^2}{\lambda^2} + \frac{m^2}{\mu^2} \ \ldots\ldots\ldots\ldots\ldots\ldots(21),$$

the quartic will have four more imaginary points of undulation lying in pairs upon DE and FG. The coordinates of the eight imaginary points can therefore be found.

Equations (17) and (21) require that $l/\lambda \pm n/\nu = 0$. If we take the upper sign it follows from (5) that $m = 0$, which is inadmissible. We must therefore take the lower sign, and we obtain from (5)

$$2l/\lambda = 2n/\nu = -m/\mu,$$

which by (17) and (21) give

$$32l^2mn = -5\lambda, \quad 16lm^2n = 5\mu, \quad 32lmn^2 = -5\nu,$$

which determine $\lambda,\ \mu,\ \nu$.

185. The above arrangement of points of undulation is not the only possible one; for the equation

$$l^4\alpha^4 + m^4\beta^4 - n^4\gamma^4 = 0$$

represents a quartic having a pair of real and a pair of imaginary points of undulation on BC and AC, and four imaginary points on AB.   Also the equation

$$\alpha\beta\gamma\,(\lambda\alpha + \mu\beta + \nu\gamma) = (l\alpha + m\beta + n\gamma)^4$$

represents a quartic having four real points of undulation on the line $(l,\ m,\ n)$.

Double Tangents.

186.   We have shown in § 180 that

$$\alpha\beta\gamma u + S^2 = 0 \dots\dots\dots\dots\dots\dots\dots(1)$$

is the equation of a quartic, four of whose double tangents are the lines α, β, γ, u; and that the points of contact are the intersections of these lines with the conic S.   Let

$$\Sigma = S + k\beta\gamma \dots\dots\dots\dots\dots\dots\dots(2)\ ;$$

then $\Sigma = 0$ is the equation of another conic which passes through the four points of contact of the double tangents β and γ with the quartic.   Substituting from (2), (1) may be written

$$\beta\gamma\,(\alpha u - 2kS - k^2\beta\gamma) + \Sigma^2 = 0 \ \dots\dots\dots\dots(3).$$

Since the terms in k cancel one another when the quartic is written out at full length, k may have any value we please; if therefore k be determined so that the discriminant of the conic in brackets vanishes, the latter will be the product of two linear factors vw, and (3) becomes

$$\beta\gamma vw + \Sigma^2 = 0 \dots\dots\dots\dots\dots\dots\dots(4).$$

We therefore obtain the theorem :—

A conic can be drawn through the eight points of contact of any four double tangents to a quartic.

The discriminant of the conic when equated to zero furnishes a quintic equation for k which involves k as a factor.   The solution $k = 0$ reproduces the conic S, whilst the four roots of the quartic factor furnish four conics of the type Σ.   Since each of these five conics passes through the four points of contact of the double tangents β and γ, it follows that :—

Through the four points of contact of any two double tangents five conics can be described, each of which passes through the four points of contact of two other double tangents.

From the table in § 158, it will be seen that all quartics having two double points, except binodal quartics, cannot have more than four double tangents. Hence the points of contact of the double tangents to all quartics, other than those of the first four species, lie on a conic.

The system of conics which can be drawn through the eight points of contact of any four double tangents has been discussed by the authorities cited below[*].

187. We have shown in § 180 that every tangent at a point of undulation is equivalent to one double and two stationary tangents; hence every anautotomic quartic which has four points of undulation has twenty-four double and sixteen stationary tangents. To find the equations of the former, we must take S as the conic inscribed in the triangle of reference, and make the discriminant of $\alpha u - 2kS - k^2\beta\gamma$ equal to zero.

The condition for this is that k should be one of the roots of (12) or (13) of § 182. Let

$$p = 4mn, \quad q^2 = \mu\nu/\lambda, \quad P = (p^2 - q^2)^{\frac{1}{2}};$$

then the roots of (12) are

$$k = p \pm P.$$

Taking the upper sign, and using (5) of § 181, the conic can be reduced to

$$\left(\frac{m^2}{\mu^2} + \frac{n^2}{\nu^2} + \frac{P}{2\mu\nu}\right)\lambda^2\alpha^2 + m^2\beta^2 + n^2\gamma^2 + \tfrac{1}{2}P\beta\gamma$$
$$+ \left(\frac{2n^2}{\nu} + \frac{P}{2\mu}\right)\gamma\alpha + \left(\frac{2m^2}{\mu} + \frac{P}{2\nu}\right)\alpha\beta = 0,$$

which splits up into the factors

$$\left\{\frac{m}{\mu} + (P \pm \iota q)\frac{n}{p\nu}\right\}\lambda\alpha + m\beta + (P \pm \iota q)\frac{n\gamma}{p} = 0.$$

If $\mu\nu/\lambda$ is positive ιq is imaginary, and the double tangents are imaginary; but if $\mu\nu/\lambda$ is negative, the double tangents are real.

The equations of the remaining double tangents can be found in a similar manner.

[*] Salmon, *Higher Plane Curves*, Chap. vi.; Hesse, *Crelle*, Vol. xlix. p. 243; Cayley, *Crelle*, lxviii. p. 176 and *Collected Papers*, Vol. vii. p. 123; Geiser, *Math. Ann.* Vol. i. p. 129; Aronhold, *Berlin. Monatsberichte*, 1864, p. 499.

Singularities at Infinity.

188. When a quartic has a singularity at infinity, the equation of the curve may be found in the manner explained in § 47, and we shall proceed to find the Cartesian equation when the singularity lies on the axis of x. To do this we take a triangle of reference whose angle B is a right angle, and suppose the singularity at A. We then take BA and BC as the axes of x and y and transform the trilinear equations given in §§ 165 and 168 by putting

$$\alpha = x, \quad \beta = 1, \quad \gamma = y,$$

for since β becomes the line at infinity, we may without loss of generality suppose it equal to unity. The equations are then as follows, where U_n, V_n denote polynomials in y of degree n.

Tacnode.

$$m^2 + 2my(\lambda x + \mu) + y^2(V_0 x^2 + x V_1 + V_2) = 0.$$

Rhamphoid-cusp.

$$(m + \lambda xy)^2 + 2m\mu y + y^2(x V_1 + V_2) = 0.$$

Oscnode.

$$(m + \lambda xy + \mu y)^2 + y^2(q xy + V_2) = 0.$$

Tacnode-cusp.

$$(m + \lambda xy + \mu y + ky^2)^2 + Ey^3 + Fy^4 = 0.$$

Flecnode.

$$x^2 U_1 V_1 + x U_1 V_2 + U_4 = 0.$$

Biflecnode.

$$x^2 U_2 + x U_2 V_1 + U_4 = 0.$$

Triple point.

$$x U_3 + U_4 = 0.$$

Point of undulation.

$$y S_3 + (xy + A y^2 + 2By + C)^2 = 0,$$

where $S_3 = 0$ is the equation of any cubic curve, and the axis of x is the tangent at the point of undulation.

With the exception of the flecnode, biflecnode and point of undulation, a quartic curve cannot have more than one singularity of the preceding character. Hence the discussion of curves having a pair of imaginary singularities of the latter kind at the circular points at infinity belongs to the theory of curves of a higher degree than the fourth. Quartic curves having nodes or cusps at the circular points will be considered in the next chapter, whilst the investigation of the equations of quartics having imaginary points of inflexion or undulation at the circular points may be left to the reader.

189. We have shown in §§ 79 and 80 that if a curve of class m has a pair of *nodes* at the circular points, and *in addition* has δ nodes and κ cusps, the curve has two double foci and $m + 2\delta + 3\kappa - 4$ single foci, which may however for certain values of the constants coalesce into one or more multiple foci. If however the curve has a pair of *flecnodes* at the circular points, the point of intersection of the two inflexional tangents will be a triple focus, and consequently the curve will have $m + 2\delta + 3\kappa - 5$ single foci, one triple and one double focus. And if the circular points are *biflecnodes*, the curve will have $m + 2\delta + 3\kappa - 6$ single foci and two triple foci. We shall hereafter show that the Cassinian, for which $m = 8$, $\delta = 0$, $\kappa = 0$, has a pair of triple foci and a pair of single foci; whilst the lemniscate, which is a particular case of the Cassinian, for which $m = 6$, $\delta = 1$, $\kappa = 0$, has a pair of triple foci and a double focus at the real biflecnode which is formed by the union of the two single foci of the Cassinian.

Binodal Quartics.*

190. The general equation of a binodal quartic whose nodes are B and C is

$$\alpha^3 u + \lambda^2 \beta^2 \gamma^2 + \mu^2 \gamma^2 \alpha^2 + \nu^2 \alpha^2 \beta^2 + \alpha\beta\gamma v = 0 \quad \ldots\ldots\ldots(1),$$

where
$$u = L^2\alpha + M\beta + N\gamma,$$
$$v = l\alpha + m\beta + n\gamma.$$

---

* The theory of anautotomic quartics has been considered by Zeuthen in a series of memoirs published in the *Mathematische Annalen*, where a variety of papers by Brill, Klein and other German mathematicians bearing on the subject will be found. Uninodal quartics have been discussed by W. R. W. Roberts, *Proc. Lond. Math. Soc.* Vol. xxv. pp. 151—172; and unicuspidal quartics by H. W. Richmond, *Quart. Journ.* Vol. xxvii. p. 5. Reference may also be made to *The Forms of Plane Quartic Curves* by Miss Gentry, published by Robert Drummond of New York; to the Index of Papers, *Proc. Lond. Math. Soc.* Vol. xxx.; to Prof. Cayley's *Collected Papers*; and to the papers of H. M. Jeffrey in the *Quarterly Journal*.

Let the lines β and γ be chosen so that they are two of the tangents to the quartic from the nodes B and C; then

$$N = 2L\mu, \quad M = 2L\nu,$$

and (1) becomes

$$L\alpha^3 (L\alpha + 2\nu\beta + 2\mu\gamma) + \lambda^2\beta^2\gamma^2 + \mu^2\gamma^2\alpha^2 + \nu^2\alpha^2\beta^2 + \alpha\beta\gamma u = 0 \ldots(2).$$

The equation of the line joining the points of contact of β and γ with the quartic is

$$L\alpha + \nu\beta + \mu\gamma = 0,$$

and (2) may be written in the form

$$\{\lambda\beta\gamma + \alpha (L\alpha + \nu\beta + \mu\gamma)\}^2$$
$$+ \alpha\beta\gamma \{(l - 2L\lambda - 2\mu\nu)\, \alpha + (m - 2\nu\lambda)\, \beta + (n - 2\lambda\mu)\, \gamma\} = 0 \ldots(3).$$

The form of this equation shows that

$$(l - 2L\lambda - 2\mu\nu)\, \alpha + (m - 2\nu\lambda)\, \beta + (n - 2\lambda\mu)\, \gamma = 0 \ \ldots\ldots(4)$$

is one of the double tangents; also since (2) is unaltered when the sign of λ is changed, another double tangent is

$$(l + 2L\lambda - 2\mu\nu)\, \alpha + (m + 2\nu\lambda)\, \beta + (n + 2\lambda\mu)\, \gamma = 0 \ldots\ldots(5).$$

Equation (2) also remains unaltered when the signs of L, μ, ν are changed; but this would merely reproduce equations (4) and (5).

Since (3) is of the form $S^2 + uvwt = 0$, the remaining six double tangents can be found by the method explained in § 180. If however the quartic has a cusp at B, $m = \pm 2\lambda\nu$; taking the upper sign, it follows that (4) is not a double tangent, but one of the tangents drawn from the cusp; and the double tangents consist of (5) and three others. If C is also a cusp, $n = 2\lambda\mu$, and the only double tangent is given by (5).

Trinodal Quartics.

191. Every trinodal quartic has four double tangents, which will however be reduced in number if any of the nodes become cusps; also since the curve is of the sixth class, only two tangents can be drawn from a node to the curve. The bitangential curve is obviously a conic.

To find the equations of the four double tangents and of the bitangential conic.*

---

* H. M. Taylor, *Proc. Lond. Math. Soc.* Vol. xxviii. p. 316.

Let the nodes be situated at the angular points of the triangle of reference; then the equation of the quartic is

$$\lambda^2\beta^2\gamma^2 + \mu^2\gamma^2\alpha^2 + \nu^2\alpha^2\beta^2 + 2\alpha\beta\gamma\,(l\alpha + m\beta + n\gamma) = 0\ \ldots\ldots(1),$$

which may be written in the form

$$(\lambda\beta\gamma + \mu\gamma\alpha + \nu\alpha\beta)^2 + 2\alpha\beta\gamma\,\{(l - \mu\nu)\,\alpha + (m - \nu\lambda)\,\beta + (n - \lambda\mu)\,\gamma\} = 0$$
$$\ldots\ldots\ldots(2),$$

which shows that the line

$$(l - \mu\nu)\,\alpha + (m - \nu\lambda)\,\beta + (n - \lambda\mu)\,\gamma = 0\ \ldots\ldots\ldots(3)$$

is a double tangent.

Since (1) remains unaltered when the sign of any one of the quantities λ, μ, ν is changed, we obtain the equations of the three other double tangents by writing $-\lambda$, $-\mu$, $-\nu$ respectively for λ, μ, ν in (3).

The equation of the conic passing through the eight points of contact of the double tangents can be shown to be

$$(l\alpha + m\beta + n\gamma)^2 - \mu^2\nu^2\alpha^2 - \nu^2\lambda^2\beta^2 - \lambda^2\mu^2\gamma^2 = 0\ \ldots\ldots\ldots(4),$$

for if we multiply the equations of the four double tangents together and subtract the square of (4), it will be found that the resulting equation reduces to (1).

When the quartic has three biflecnodes, $l = m = n = 0$, and the conic (4) is self-conjugate to the triangle formed by joining the three nodes; and when the quartic is tricuspidal, the coefficients of α^2, β^2, γ^2 vanish, and (4) becomes a conic circumscribing the triangle in question.

192.　We shall add a few miscellaneous propositions concerning trinodal quartics.

The six nodal tangents to a trinodal quartic touch a conic.

From (1) it appears that the equation of the nodal tangents at A is

$$\nu^2\beta^2 + \mu^2\gamma^2 + 2l\beta\gamma = 0\ \ldots\ldots\ldots\ldots\ldots\ldots(5),$$

which may be written in the form

$$(\eta_1\beta + \zeta_1\gamma)\,(\eta_2\beta + \zeta_2\gamma) = 0,$$

where $\qquad \dfrac{\zeta_1\zeta_2}{\eta_1\eta_2} = \dfrac{\mu^2}{\nu^2}\,;\ \ \dfrac{\zeta_1}{\eta_1} + \dfrac{\zeta_2}{\eta_2} = \dfrac{2l}{\nu^2}\ \ \ldots\ldots\ldots\ldots(6).$

The tangential equation of a conic is

$$P\xi^2 + Q\eta^2 + R\zeta^2 + 2p\eta\zeta + 2q\zeta\xi + 2r\xi\eta = 0\ldots\ldots\ldots(7),$$

where $(\xi,\ \eta,\ \zeta)$ are tangential coordinates. To find the condition that (5) should touch (7), put $\xi = 0$, and (7) becomes

$$Q\eta^2 + R\zeta^2 + 2p\eta\zeta = 0,$$

whence
$$\frac{\zeta_1\zeta_2}{\eta_1\eta_2} = \frac{Q}{R} = \frac{\mu^2}{\nu^2},$$

$$\frac{\zeta_1}{\eta_1} + \frac{\zeta_2}{\eta_2} = -\frac{2p}{R} = \frac{2l}{\nu^2},$$

by (6). Hence if we take

$$P/\lambda^2 = Q/\mu^2 = R/\nu^2 = -p/l = -q/m = -r/n,$$

the conic (7) becomes

$$\lambda^2\xi^2 + \mu^2\eta^2 + \nu^2\zeta^2 - 2l\eta\zeta - 2m\zeta\xi - 2n\xi\eta = 0\ldots\ldots\ldots(8),$$

which is the equation of a conic touching the six nodal tangents. By § 71 equation (8) when expressed in trilinear coordinates becomes

$$(\mu^2\nu^2 - l^2)\,\alpha^2 + (\nu^2\lambda^2 - m^2)\,\beta^2 + (\lambda^2\mu^2 - n^2)\,\gamma^2$$
$$+ 2\,(mn + \lambda^2 l)\,\beta\gamma + 2\,(nl + \mu^2 m)\,\gamma\alpha + 2\,(lm + \nu^2 n)\,\alpha\beta = 0\ldots(9).$$

Equation (9) may also be expressed in the form

$$\nu^2\beta^2 + \mu^2\gamma^2 + 2l\beta\gamma + k^2\,\{(\mu^2\nu^2 - l^2)\,\alpha + (lm + \nu^2 n)\,\beta + (ln + \mu^2 m)\,\gamma\}^2 = 0,$$

where

$$k^2\,(\lambda^2\mu^2\nu^2 - l^2\lambda^2 - m^2\mu^2 - n^2\nu^2 - 2lmn) = 1,$$

which shows that the term in brackets is the chord of contact. The equations of the other chords of contact can be obtained in a similar manner.

When the nodes are biflecnodes, $l = m = n = 0$; and the conic is self-conjugate to the nodal triangle, and becomes identical with (4).

When the three double points are cusps, $l = \mu\nu$ &c., and the coefficients of α^2, β^2, γ^2 vanish. This requires that $\lambda^2 l = mn$ &c.; whence the curve becomes

$$\beta\gamma/l + \gamma\alpha/m + \alpha\beta/n = 0,$$

which represents a conic circumscribing the nodal triangle.

193. *From each node of a trinodal quartic two tangents can be drawn to the curve, and these six tangents touch a conic.*

Let $\beta = k\gamma$ be one of the tangents drawn to the quartic from the node A. Substitute in (1), divide out by γ^2 and express the condition that the resulting quadratic in α/γ should have equal roots; this gives a quadratic equation for k, and on substituting β/γ for k we obtain

$$(\lambda^2\nu^2 - m^2)\,\beta^2 + 2\,(l\lambda^2 - mn)\,\beta\gamma + (\lambda^2\mu^2 - n^2)\,\gamma^2 = 0,$$

which is the equation of the two tangents drawn from the node A.

Let $\quad \sigma_1 = \mu^2\nu^2 - l^2, \quad \sigma_2 = \nu^2\lambda^2 - m^2, \quad \sigma_3 = \lambda^2\mu^2 - n^2;$

then proceeding as in § 192 we shall find that the tangential equation of the conic which touches the six tangents is

$$\sigma_2\sigma_3\xi^2 + \sigma_3\sigma_1\eta^2 + \sigma_1\sigma_2\zeta^2 + 2\sigma_1\,(mn - l\lambda^2)\,\eta\zeta$$
$$+ 2\sigma_2\,(nl - m\mu^2)\,\zeta\xi + 2\sigma_3\,(lm - n\nu^2)\,\xi\eta = 0,$$

and the trilinear equation is

$$\sigma_1^2\lambda^2\alpha^2 + \sigma_2^2\mu^2\beta^2 + \sigma_3^2\nu^2\gamma^2 + 2l\sigma_2\sigma_3\beta\gamma + 2m\sigma_3\sigma_1\gamma\alpha + 2n\sigma_1\sigma_2\alpha\beta = 0.$$

194. The following additional properties of trinodal quartics may be mentioned[*].

(i) *The six points of inflexion lie on a conic.*

(ii) *The six points of contact of the tangents drawn from the nodes lie on a second conic.*

(iii) *The six points in which the nodal tangents intersect the quartic lie on a third conic.*

(iv) *The three conics pass through two points P and Q on the quartic, which lie on the conic*

$$\lambda^2 l\beta\gamma + \mu^2 m\gamma\alpha + \nu^2 n\alpha\beta = 0.$$

We shall prove the third theorem as an example of the mode of dealing with such questions.

If in (1) we choose three new coordinates α', β', γ' such that $\alpha/\lambda = \alpha'$, &c., and then change the constants l, m, n; the equation of a trinodal quartic may be written in the form

$$\alpha^2\,(\beta^2 + \gamma^2 + l\beta\gamma) + \beta\gamma\,\{\beta\gamma + \alpha\,(m\beta + n\gamma)\} = 0,$$

---

[*] Brill, *Math. Annalen*, Vol. XII. p. 90; XIII. p. 175; F. Meyer, *Apolarität und Rationale Curven*, pp. 283—7.

which shows that the nodal tangents at A, whose equation is

$$\beta^2 + \gamma^2 + l\beta\gamma = 0 \dots\dots\dots\dots\dots\dots(10),$$

intersect the quartic at the two points D, D' where it is cut by the conic

$$\beta\gamma + \alpha(m\beta + n\gamma) = 0 \dots\dots\dots\dots\dots(11).$$

This conic circumscribes the triangle of reference, and therefore passes through the three nodes which make up the remaining six points of intersection of the conic and the quartic.

From (10) and (11) it can be shown that the equation of the line DD' is

$$k_1\alpha + \beta/m + \gamma/n = 0 \dots\dots\dots\dots\dots(12),$$

where
$$k_1 = m/n + n/m - l \dots\dots\dots\dots\dots\dots(13).$$

By cyclical interchanges of the letters (α, β, γ) and (l, m, n) the corresponding results for the nodal tangents at B and C and the corresponding points of intersection E, E' and F, F' can be obtained.

The equation of the quartic may also be written in the form

$$(\beta^2 + \gamma^2 + l\beta\gamma)(\gamma^2 + \alpha^2 + m\gamma\alpha) - \gamma^2\{\gamma^2 + l\beta\gamma + m\gamma\alpha + (lm - n)\alpha\beta\} = 0$$
$$\dots\dots(14),$$

the first term of which is the product of the equations of the nodal tangents at A and B. The form of (14) shows that the conic

$$\gamma^2 + l\beta\gamma + m\gamma\alpha + (lm - n)\alpha\beta = 0 \dots\dots\dots\dots(15)$$

passes through the points of intersection D, D' and E, E' of the nodal tangents at A and B respectively; accordingly we obtain the following theorem :—

A conic can be described through any two nodes of a trinodal quartic and the four points at which the tangents at these nodes intersect the quartic.

Let
$$S = (P, Q, R, P', Q', R'\!\!\times\!\alpha, \beta, \gamma)^2 = 0 \dots\dots\dots(16)$$

be the equation of the proposed conic which is assumed to pass through the six points D, D'; E, E'; F, F' in which the nodal tangents at A, B and C intersect the quartic. Then the equation

$$S + (k_1\alpha + \beta/m + \gamma/n)(\alpha/l + k_2\beta + \gamma/n) = 0\dots\dots\dots(17)$$

represents any conic which passes through D, D'; E, E'. This conic may therefore be made to represent (15), in which case we must have

$$P = -k_1/l, \quad Q = -k_2/m \dots\dots\dots\dots(18),$$

$$R + 1/n^2 = (2P' + k_2/n + 1/mn)/l = (2Q' + k_1/n + 1/ln)/m$$
$$= (2R' + k_1k_2 + 1/lm)/(lm - n)\dots\dots(19).$$

If the conic passes through the two points F, F' it must be possible to make the equation

$$\mu S + (\alpha/l + k_2\beta + \gamma/n)(\alpha/l + \beta/m + k_3\gamma) = 0 \ \dots\dots(20)$$

represent the conic

$$\alpha^2 + (mn - l)\,\beta\gamma + m\gamma\alpha + n\alpha\beta = 0 \ \dots\dots\dots\dots(21),$$

which by virtue of (15) is the conic which passes through E, E'; F, F'; B, C. Comparing (20) and (21) we obtain

$$\mu = 1, \quad Q = -k_2/m, \quad R = -k_3/n \ \dots\dots\dots\dots(22),$$

$$P + 1/l^2 = (2P' + k_2k_3 + 1/mn)/(mn - l) = (2Q' + k_3/l + 1/ln)/m$$
$$= (2R' + k_2/l + 1/lm)/n \ \dots\dots\dots\dots(23).$$

Equations (19) and (23) are *six* equations for determining three quantities P', Q', R'; but on solving them it will be found that they are capable of coexisting, which shows that a conic S can be described through the six points D, D'; E, E'; F, F'. The values of P, Q, R are determined by (18) and (22), and by solving (19) and (23) and taking account of the values of k_1, k_2, k_3 determined by (13), we shall obtain

$$2P' = l - \frac{1}{l} - \frac{l^2 + 1}{mn} \ \dots\dots\dots\dots(24),$$

with symmetrical expressions for Q', R'. The conic is therefore completely determined by (16), (18), (22), and (24).

195. Since a real crunode reduces the number of real points of inflexion by two and the number of imaginary ones by four, whilst a conjugate point or an imaginary node reduces the number of imaginary points of inflexion by six, the number of real and imaginary points of inflexion of any *given* trinodal quartic can be written down. The same can also be done in the case of quartics having three double points, some of which are cusps.

9—2

196.   A tricuspidal quartic is the reciprocal polar of a nodal cubic, from which it follows (i) that the three cuspidal tangents intersect at a point; (ii) that such a quartic has only one double tangent, which must be real; (iii) that its points of contact are real when two of the cusps are imaginary, and imaginary when all three cusps are real.   It follows from (1) and (2) that when all three cusps are real

$$l = \pm \mu\nu, \quad m = \pm \nu\lambda, \quad n = \pm \lambda\mu,$$

whence the cuspidal tangents are

$$\mu\gamma = \nu\beta, \quad \nu\alpha = \lambda\gamma, \quad \lambda\beta = \mu\alpha,$$

which meet at the point

$$\alpha/\lambda = \beta/\mu = \gamma/\nu,$$

whilst the quartic is reducible to the form

$$\pm (\lambda/\alpha)^{\frac{1}{2}} \pm (\mu/\beta)^{\frac{1}{2}} \pm (\nu/\gamma)^{\frac{1}{2}} = 0.$$

The two most interesting quartics of this species are the cardioid and the three-cusped hypocycloid, whose properties will be discussed in Chapters X. and XI.   It will further be shown in Chapter XII. that any tricuspidal quartic can be projected into either of these curves.   Hence a detailed discussion of tricuspidal quartics is unnecessary, since all their projective properties can be deduced from the known properties of the above-mentioned two curves.

CHAPTER IX.

BICIRCULAR QUARTICS.

197. A CLASS of quartics, which include a variety of well known curves, possesses a pair of nodes or a pair of cusps at the circular points at infinity. The former class belongs to species IV., VII. or VIII. and are called *bicircular quartics*; and the latter to species VI., IX. or X., and are called *cartesians* because the oval of Descartes was one of the first curves of this kind which was studied.

198. *To find the equation of a bicircular quartic.*

The general equation in trilinear coordinates of a quartic having a pair of nodes at B and C is

$$\alpha^3 (L\alpha + M\beta + N\gamma) + \lambda\beta^2\gamma^2 + \mu\gamma^2\alpha^2 + \nu\alpha^2\beta^2 + \alpha\beta\gamma (l\alpha + m\beta + n\gamma) = 0 \qquad \ldots\ldots\ldots(1).$$

In this equation the quantities α, β, γ may be any real or imaginary straight lines. If, therefore, we suppose that B and C are the circular points at infinity and that A is the origin of a pair of rectangular axes, we can transform (1) into Cartesian coordinates by putting

$$\alpha = I, \quad \beta = x + \iota y, \quad \gamma = x - \iota y \quad \ldots\ldots\ldots\ldots(2),$$

and (1) becomes

$$I^3 \{LI + (M + N) x + \iota (M - N) y\} + \lambda (x^2 + y^2)^2$$
$$+ I^2 \{(\mu + \nu) (x^2 - y^2) + 2\iota (\mu - \nu) xy\}$$
$$+ I (x^2 + y^2) \{lI + (m + n) x + \iota (m - n) y\} = 0 \ldots\ldots(3).$$

Changing the constants so that $\iota (M - N)$, $\iota (\mu - \nu)$, $\iota (m - n)$ are represented by real quantities, (3) may be written in the form

$$(x^2 + y^2)^2 u_0 + (x^2 + y^2) u_1 + v_2 + v_1 + v_0 = 0 \quad \ldots\ldots\ldots(4),$$

where u_n, v_n are binary quantics in x and y. Equation (4) may also be written in the form

$$S^2 + U = 0 \quad \ldots\ldots\ldots\ldots\ldots\ldots(5),$$

where S is a circle and U a conic; or in the form

$$S^2 + UI^2 = 0 \quad \ldots\ldots\ldots\ldots\ldots\ldots(6),$$

where S and U have the same meanings and I is the line at infinity.

Equation (6) shows that the conic U and the line at infinity I have a contact of the first order with the quartic at the points where it is cut by the circle S; and that this circle also has a contact of the first order with the quartic at the two points where it intersects the line at infinity. The conic U *touches* the quartic at the four points where S and U intersect; but the contact of the circle and the line at infinity with the quartic arises from the fact that both pass through the circular points, which are nodes on the quartic.

199. *To find the equation of a cartesian.*

The equation

$$\alpha^3\,(L\alpha + M\beta + N\gamma) + \alpha^2\,(\lambda^2\beta^2 + 2\mu\beta\gamma + \nu^2\gamma^2) + 2k\iota\beta\gamma\,(\lambda\beta + \nu\gamma)$$
$$+ k^2\beta^2\gamma^2 = 0 \quad \ldots\ldots\ldots\ldots(7)$$

represents a quartic having a pair of cusps at B and C.

Transform this equation by means of (2) and then put

$$\lambda + \nu = p, \quad \iota\,(\lambda - \nu) = q,$$

and it becomes

$$k^2\,(x^2 + y^2)^2 + 2kI\,(x^2 + y^2)\,(px + qy) + I^2\left\{\tfrac{1}{2}\,(p^2 - q^2)\,(x^2 - y^2)\right.$$
$$\left. + 2pqxy + 2\mu\,(x^2 + y^2)\right\} + I^3\,(LI + Px + Qy) = 0.$$

Let

$$Ip/k = a, \quad Iq/k = b, \quad I^2\left\{2\mu - \tfrac{1}{2}\,(p^2 + q^2)\right\}/k^2 = 2c^2,$$

and the equation may be written in the form

$$(x^2 + y^2 + ax + by)^2 + 2c^2\,(x^2 + y^2) + I^3\,(LI + Px + Qy)/k^2 = 0,$$

which is the same as

$$(x^2 + y^2 + ax + by + c^2)^2 + A\,x + B\,y + C = 0.$$

The equation of a cartesian may therefore be written in either of the forms

$$S^2 + u = 0 \quad\dots\dots\dots\dots\dots\dots\dots(8),$$

or
$$S^2 + I^3 u = 0 \quad\dots\dots\dots\dots\dots\dots(9),$$

where S is a circle and u is a straight line. The form of (9) shows that the line $u = 0$ is the only double tangent which the curve can have; and also that the circle S has a contact of the second order with the curve at each of the circular points at infinity.

200. Bicircular quartic curves have formed the subject of an exhaustive memoir by the late Dr Casey*, from which most of the present chapter will be taken. He first of all shows that the quartic may be generated in the following manner :—

If OT be the perpendicular from any fixed point O on to the tangent at any point Q of a fixed conic; and if two points P, P' be taken on OT such that TP = TP', and

$$OT^2 - PT^2 = \delta^2\dots\dots\dots\dots\dots\dots(10),$$

where δ is a constant, the locus of P and P' is a bicircular quartic.

When the fixed conic is an ellipse or hyperbola, the quartic has two nodes at the circular points at infinity; when the conic is a circle, the circular points are cusps and the quartic is a cartesian; and when the conic is a parabola, the curve degenerates into a circular cubic.

Let EY be the perpendicular from the centre E of the conic on to the tangent at any point Q. Let (f, g) be the coordinates of O referred to E; (x, y) those of P referred to O. Let $OP = r$, $EY = p$, $YEX = \phi$. Then

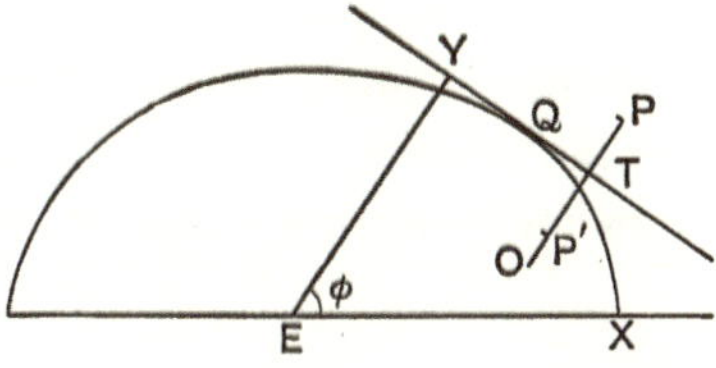

$$OT = p - f \cos \phi - g \sin \phi,$$
$$PT = r - OT,$$

* *Trans. R. I. A.* Vol. xxiv. p. 457.

also by (10)

$$\delta^2 = 2r \,.\, OT - r^2,$$

whence

$$r^2 + 2fx + 2gy + \delta^2 = 2rp \quad \ldots\ldots\ldots\ldots(11).$$

Now

$$p^2 = a^2 \cos^2 \phi + b^2 \sin^2 \phi,$$

whence (11) becomes

$$(r^2 + 2fx + 2gy + \delta^2)^2 = 4(a^2x^2 + b^2y^2) \quad \ldots\ldots\ldots(12),$$

which by (5) is the equation of a bicircular quartic.

When the conic is a circle, $a = b$, and (12) may be put into the form

$$(r^2 + 2fx + 2gy + \delta^2 - 2a^2)^2 = 4a^2(a^2 - 2fx - 2gy - \delta^2) \ldots(13),$$

which is the equation of a cartesian.

When the conic is a parabola whose focus is E and vertex X, $p = a \sec \phi$, and (11) becomes

$$x(r^2 + 2fx + 2gy + \delta^2) = 2ar^2 \quad \ldots\ldots\ldots\ldots(14),$$

which is the equation of a circular cubic.

The fixed conic is called the *focal conic* because, as will be shown hereafter, it passes through four of the foci of the quartic.

If the quartic (12) be inverted from O with respect to a circle of radius δ, it is inverted into itself. Hence O is called a *centre of inversion*, and the circle whose centre is O and radius δ is called a *circle of inversion*. We shall hereafter prove that, in general, a bicircular quartic has four centres and four circles of inversion.

Equation (12) contains five independent constants; and if the origin be transferred to any arbitrary point and the axes be turned through any arbitrary angle, three more constants will be introduced. Hence the general equation of a bicircular quartic contains eight constants, and that of a cartesian seven.

201. *The inverse of a bicircular quartic is another bicircular quartic unless the centre of inversion lies on the curve, in which case it is a circular cubic.*

The general equation is of the form

$$r^4 v_0 + r^2 v_1 + u_2 + u_1 + u_0 = 0 \quad \ldots\ldots\ldots\ldots(15),$$

the inverse of which with respect to the origin is obviously
a curve of the same form. If, however, the origin is situated
on the curve, $u_0 = 0$, in which case the inverse curve reduces to
a circular cubic.

202. Bicircular quartics and cartesians may be divided into
two classes according as the curve has two or three double points.
In the latter case the curve is the inverse and also the pedal of
a conic with respect to some point in its plane, which is the
third double point of the quartic. That a bicircular quartic
having three double points is the inverse of a conic, can be at
once shown by taking the third double point as the origin, in
which case (15) reduces to $r^4 v_0 + r^2 v_1 + u_2 = 0$, the inverse of which
is a conic. We shall now prove that :—

*The inverse of a conic with respect to any point not on the
curve is a bicircular quartic having a third double point at the
centre of inversion; and this point will be a node, a cusp or a
conjugate point according as the conic is a hyperbola, a parabola
or an ellipse.*

The equation of a central conic referred to any point (f, g) as
origin is

$$\frac{(x+f)^2}{a^2} + \frac{(y+g)^2}{b^2} = 1,$$

the inverse of which is

$$k^4 \left(\frac{x^2}{a^2} + \frac{y^2}{b^2}\right) + 2k^2 r^2 \left(\frac{fx}{a^2} + \frac{gy}{b^2}\right) + \left(\frac{f^2}{a^2} + \frac{g^2}{b^2} - 1\right) r^4 = 0 \quad \ldots(16),$$

and the origin will therefore be a node or a conjugate point
according as the conic is a hyperbola or an ellipse. When the
conic is a parabola, the equation of the curve is

$$(y+g)^2 = 4a(x+f)$$

and the inverse curve is

$$k^4 y^2 + 2k^2 r^2 (gy - 2ax) + (g^2 - 4af) r^4 = 0 \quad \ldots\ldots(17),$$

and the origin is a cusp.

203. When the centre of inversion is the focus of the conic
the quartic becomes a cartesian, which is called a limaçon when
the conic is an ellipse or hyperbola, and a cardioid when the
conic is a parabola. When the centre of inversion lies on the
curve, the quartic degenerates into a circular cubic. We have
also shown in § 170 that the lemniscate of Bernoulli is the only

trinodal quartic which possesses a pair of biflecnodes at the circular points, in which case the conic is a rectangular hyperbola and the centre of inversion is the centre of the hyperbola. It only remains therefore to consider the case in which the quartic has a pair of *flecnodes* at the circular points.

Equation (5) of § 169 is the equation of a trinodal quartic having a pair of flecnodes at B and C. Putting $\alpha = I$, $\beta = x + \iota y$, $\gamma = x - \iota y$, it will be found that in order that the resulting curve should be real, we must have $l(p + q) = 1$; whence, putting $I/nq = A$, $lq = B$, the equation of the curve becomes

$$(x^2 + y^2)^2 + 2Ax(x^2 + y^2) + A^2B\{(3 - 2B)x^2 - (1 - 2B)y^2\} = 0,$$

which is the equation of a bicircular quartic having a pair of flecnodes at the circular points. The origin will be a cusp when $2B = 3$; but if $2B = 1$, the curve degenerates into the square of a circle.

Comparing the last equation with (16), we find that the centre of inversion is given by the equations

$$f^2(a^2 - b^2) = (a^2 + b^2)^2, \quad g = 0,$$

hence this point is determined by the following construction. *From either focus draw an ordinate cutting the director circle in P, and let the tangent at P intersect the transverse axis of the conic in T, then T is the required point.*

When the conic is a parabola, the equation of the quartic is

$$(x^2 + y^2)^2 + 2Ax(x^2 + y^2) + 3A^2y^2 = 0,$$

and the point T lies on the opposite side of the directrix at a distance equal to that of the focus.

204. *The pedal of a central conic with respect to any point in its plane is a bicircular quartic having a third double point at the origin, which is a node, a cusp or a conjugate point according as the origin lies without, upon or within the conic; but the pedal of a parabola is a circular cubic.*

The pedal of a central conic with respect to any origin, whose coordinates with respect to the centre are (f, g), is

$$(r^2 + fx + gy)^2 = a^2x^2 + b^2y^2.$$

The origin will accordingly be a node, a cusp or a conjugate point according as

$$f^2/a^2 + g^2/b^2 > \text{ or } = \text{ or } < 1.$$

The pedal of a parabola is

$$(r^2 + gy + fx)\, x = ay^2,$$

which is a circular cubic.

We also observe that in both these cases $\delta = 0$.

205. The preceding methods are not the only ones by which a bicircular quartic can be generated. We shall now show that:—

A bicircular quartic is the envelope of a variable circle whose centre moves along a fixed conic, called the focal conic, and which cuts a fixed circle orthogonally.

In the figure to § 200, describe a circle whose centre is Q and which passes through P and P'. Then, since (10) may be written in the form $OP . OP' = \delta^2$, it follows that the tangent from O to this circle is constant and equal to the radius δ of the fixed circle. Hence, if with O as a centre a circle of radius δ be described, this circle will cut the circle through QPP' orthogonally.

Let Q' be a point on the conic near Q; then Q' may be regarded as lying on the tangent at Q. Hence, if a circle be described through $Q'PP'$, PP' will be the radical axis of the two circles, and both will be cut orthogonally by the fixed circle. Hence P and P' will be the limiting positions of the points of intersection of the two circles, and therefore the quartic is the envelope of the moving circle.

The moving circle is called the *generating circle*; whilst by § 200 the fixed circle is the *circle of inversion*.

206. *If through the centre of inversion O any chord be drawn and P and P' be the two inverse points of intersection, the locus of the points of intersection of the normals to the quartic at P and P' is the focal conic.*

Let $OT = p$, $OP = r$, ψ the angle which the tangent to the quartic at P makes with OP; then

$$\tan \psi = r\,\frac{d\phi}{dr} = \frac{r}{p}\,\frac{dp}{dr} \cdot p\,\frac{d\phi}{dp}$$

$$= \frac{r}{p}\,\frac{dp}{dr}\,\tan OQT = \frac{r}{QT}\,\frac{dp}{dr}\,.$$

But $$2rp - r^2 = \delta^2,$$

whence $$\frac{dp}{dr} = \frac{PT}{r}.$$

Accordingly $$\tan \psi = PT/QT = \tan PQT,$$

whence PQ and $P'Q$ are the normals at P and P'.

207. *If a chord be drawn from a centre of inversion to meet the quartic in P and P', the locus of the point of intersection of the tangents at P and P' is a trinodal quartic, having three biflecnodes at the angular points of the triangle which is self-conjugate to the circle of inversion and the corresponding focal conic.*

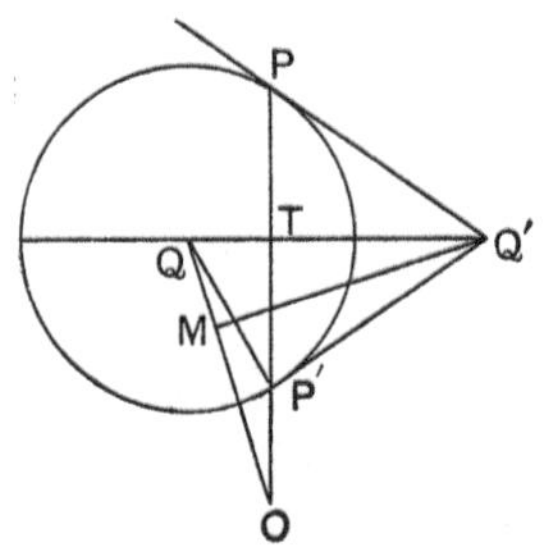

From the last proposition, it follows that since QP, QP' are the normals to the quartic at P and P', the tangents at these points are also the tangents to the generating circle whose centre is Q, and will therefore intersect at a point Q' which lies on the tangent at Q to the focal conic. Draw $Q'M$ perpendicular to OM; then since the points $Q'PQMP'$ lie on a circle,

$$OM \cdot OQ = OP \cdot OP' = \delta^2,$$

and therefore $Q'M$ is the polar of Q with respect to the circle of inversion δ.

Let the circle of inversion and the focal conic be referred to their common self-conjugate triangle; and let $(\xi,\ \eta,\ \zeta)$ be the coordinates at Q. The equation of the circle of inversion is

$$a\alpha^2 \cos A + b\beta^2 \cos B + c\gamma^2 \cos C = 0,$$

and that of the focal conic is

$$\lambda\alpha^2 + \mu\beta^2 + \nu\gamma^2 = 0.$$

Since $Q'M$ is the polar of Q with respect to the circle of inversion, its equation is

$$a\alpha\xi \cos A + b\beta\eta \cos B + c\gamma\zeta \cos C = 0,$$

and the equation of QQ' is

$$\lambda\alpha\xi + \mu\beta\eta + \nu\gamma\zeta = 0.$$

Eliminating (ξ, η, ζ) by means of the equation of the focal conic, the locus of Q' is of the form

$$P/\alpha^2 + Q/\beta^2 + R/\gamma^2 = 0,$$

which is the equation of a quartic having three biflecnodes.

208. *Every bicircular quartic can be expressed in the form of a ternary quadric of U, V, W, where these quantities are the equations of three circles.*

By means of a linear transformation any ternary quadric can be reduced to the sum of three squares; hence the equation in question may be written in the form

$$lU^2 + mV^2 + nW^2 = 0 \quad\ldots\ldots\ldots\ldots\ldots(18).$$

Now $U = r^2 + u_1 + u_0$ &c.; whence substituting in (18) it will be found that the equation reduces to (4).

209. We shall now examine the relations of the fixed circle to the focal conic.

The equation

$$\lambda U + \mu V + \nu W = 0 \ldots\ldots\ldots\ldots\ldots(19)$$

obviously represents a circle; and it can be shown by the usual methods that (18) is the envelope of (19), where (λ, μ, ν) are subject to the condition

$$\lambda^2/l + \mu^2/m + \nu^2/n = 0 \ldots\ldots\ldots\ldots\ldots(20).$$

Let the vertices A, B, C of the triangle of reference be the centres of U, V, W; then the distances of their centres from BC are $b \sin C, 0, 0$; whence the distance of the centre of (19) from BC is $\lambda b \sin C/(\lambda + \mu + \nu)$. Accordingly if α, β, γ be the trilinear coordinates of the centre of (19)

$$a\alpha/\lambda = b\beta/\mu = c\gamma/\nu \ldots\ldots\ldots\ldots\ldots(21).$$

Substituting in (20) it follows that the centre of (19) lies on the curve

$$a^2\alpha^2/l + b^2\beta^2/m + c^2\gamma^2/n = 0 \ldots\ldots\ldots\ldots\ldots(22),$$

which is a conic to which the triangle whose vertices are the centres of U, V, W is self-conjugate.

210. We shall now prove that the circle which cuts U, V and W orthogonally, cuts (19) orthogonally.

It is known from the geometry of the circle, that the radical axes of any three circles intersect in a point which is called the radical centre of the three circles; and that the tangents drawn from the radical centre to each of the three circles are equal. Hence the circle whose centre is the radical centre and whose radius is equal to any one of the tangents to the three circles cuts each of them orthogonally; also if any number of circles have a common radical centre, a circle can be described cutting each of them orthogonally.

Let S be the circle circumscribing the triangle of reference; then we may write

$$U = S + (l_1\alpha + m_1\beta + n_1\gamma)\, I,$$

with similar expressions for V and W. Whence the radical axes of U and V, V and W, W and U are

$$(l_1 - l_2)\,\alpha + (m_1 - m_2)\,\beta + (n_1 - n_2)\,\gamma = 0, \text{ &c., &c.}$$

The radical axis of U and (19) is

$$\{\mu\,(l_1 - l_2) - \nu\,(l_3 - l_1)\}\,\alpha + \{\mu\,(m_1 - m_2) - \nu\,(m_3 - m_1)\}\,\beta$$
$$+ \{\mu\,(n_1 - n_2) - \nu\,(n_3 - n_1)\}\,\gamma = 0,$$

which obviously passes through the radical centre of U, V and W. Hence the circle which cuts U, V, W orthogonally cuts (19) orthogonally. This circle is therefore the circle of inversion, the circle (19) is the generating circle, whilst the conic (22) is the focal conic.

211. It is shown in treatises on Conics, that if a circle and a conic intersect in four points P, Q, R, S; and if SP, RQ intersect in A; PR, QS in B; and PQ, SR in C; the triangle ABC is self-conjugate to the conic and the circle, and the orthocentre of ABC is the centre of the circle. If therefore the radii of the circles U, V, W be chosen so that the orthocentre is their radical centre, the circle through P, Q, R, S will cut (19) orthogonally. Accordingly the former circle is the fixed circle or circle of inversion, whilst (19) is the generating circle; hence the quartic may be generated in a third manner:—

Let the focal conic cut the circle of inversion in P, Q, R, S; let SP, QR intersect in A; PR, SQ in B; PQ, SR in C. With A, B, C as centres describe three circles U, V, W, whose radii are such that the orthocentre of ABC is their radical centre; then the quartic is the envelope of a variable circle whose centre lies on the focal conic and which cuts the circle of inversion and also the three circles U, V, W orthogonally, and its equation is

$$l U^2 + m V^2 + n W^2 = 0.$$

We have shown in § 75 that a focus of a curve may be regarded as an indefinitely small circle which has a double contact with the curve; from which it follows that the four points P, Q, R, S in which the focal conic intersects the circle of inversion are foci of the quartic. For this reason the conic in question is called the *focal conic*.

212. *When the circle of inversion touches its corresponding focal conic, the point of contact is a node on the quartic; and when it osculates the focal conic, the point of contact is a cusp.*

Let the circle and the focal conic touch at R; let p be the perpendicular from E the centre of the conic on to the tangent at R; ED the diameter conjugate to ER; ψ the angle which the normal at R makes with the major axis of the conic. Also let (ξ, η) be the coordinates of R referred to E.

The equation of the quartic referred to O as origin is given by (12). If therefore we transfer the origin to R, and recollect that

$$\xi = f + \delta \cos \psi, \quad \eta = g + \delta \sin \psi, \quad f \cos \psi + g \sin \psi = p - \delta,$$

$$a^2 \cos \psi = p\xi, \quad b^2 \sin \psi = p\eta,$$

(12) becomes

$$(r^2 + 2x\xi + 2y\eta + 2p\delta)^2 = 4 \left\{ a^2 x^2 + b^2 y^2 + 2p\delta (x\xi + y\eta) + p^2\delta^2 \right\}.$$

The terms of lowest dimensions are

$$x^2 (\xi^2 + p\delta - a^2) + y^2 (\eta^2 + p\delta - b^2) + 2xy \xi\eta,$$

and consequently the point of contact is a double point. The condition that this should be a node, a cusp or a conjugate point will be found on reduction to be

$$ED^2 > \text{or} = \text{or} < p\delta.$$

Now when $ED^2 = p\delta$, δ is the radius of curvature at the origin R; whence the point of contact will be a cusp when the circle of inversion osculates the focal conic.

The point of contact R must obviously be a real point, otherwise the quartic would have three imaginary double points; hence the two foci which coincide at R must be real foci. We thus obtain the following theorem :—

When a bicircular quartic has a real node, the latter arises from the union of two real single foci; and when it has a real cusp, the latter arises from the union of three real single foci. We shall have examples of this in the case of the limaçon and the cardioid.

When the circle of inversion has a double contact with the focal conic, each point of contact will be a double point on the quartic, which together with the circular points at infinity make four double points. Since this is greater than the maximum number, the quartic must break up into two conics each of which passes through the circular points at infinity, and must therefore be circles.

213. Before proceeding further with the theory of bicircular quartics, it will be desirable to consider certain geometrical propositions connected with the circle.

Let ABC be any triangle, O its orthocentre ; then

(i) *The triangle formed by joining any three of the four points A, B, C and O has the fourth point for its orthocentre.*

(ii) *The four triangles thus formed have a common nine-point circle.*

For the points D, E, F are the feet of the perpendiculars drawn from the angles of each of the four triangles on to the opposite sides; and the nine-point circle is the circle circumscribing the triangle DEF.

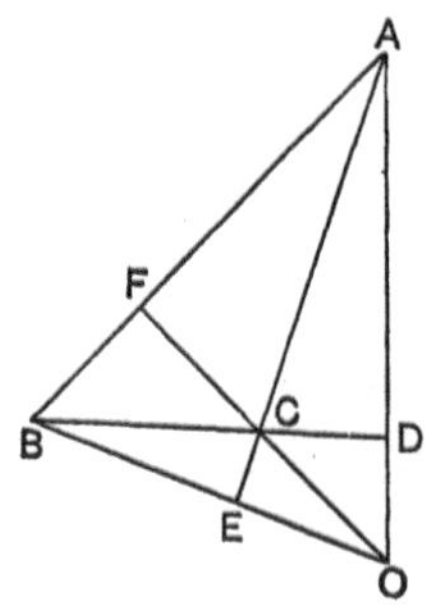

(iii) *Each point is the centre of the circle to which the triangle formed by joining the remaining three is self-conjugate.*

(iv) *The four circles, to which the four triangles are self-conjugate, cut one another orthogonally.*

Let δ, r_1, r_2, r_3 be the radii of the four circles to which the triangles ABC, OBC, OCA and OAB are respectively self-conjugate. Then since A is the pole of BC with respect to the circle to which the triangle ABC is self-conjugate,

$$\delta^2 = OD \cdot OA = -4R^2 \cos A \cos B \cos C,$$

where R is the radius of the circle circumscribing ABC. Similarly

$$r_1{}^2 = 4R^2 \cos A \sin B \sin C,$$
$$r_2{}^2 = 4R^2 \cos B \sin C \sin A,$$
$$r_3{}^2 = 4R^2 \cos C \sin A \sin B.$$

But $\qquad OA^2 - \delta^2 = 4R^2 \cos A \, (\cos A + \cos B \cos C)$
$$= 4R^2 \cos A \sin B \sin C = r_1{}^2.$$

Similarly $\qquad\qquad OB^2 - \delta^2 = r_2{}^2,$
$$OC^2 - \delta^2 = r_3{}^2,$$

which shows that the circle whose radius is δ cuts each of the circles whose radii are r_1, r_2, r_3 orthogonally. In the same way it can be proved that each of the other four circles cuts every other one orthogonally.

(v) *The radical axis of any two of the four circles passes through the centres of the remaining two.*

The radical axis of two circles is perpendicular to the line joining their centres; also since the tangents to the two circles from any point on the radical axis are equal, it follows that if from any point on the radical axis as centre a circle be described whose radius is equal to the tangent from this point to either of the two circles, the last-mentioned circle will cut the first two orthogonally. Hence the radical axis of the circles whose centres are A and B passes through the points O and C.

214. *To find the equations of the four circles.*

Let $S = 0$, $U = 0$, $V = 0$, $W = 0$ be the equations of the four orthogonal circles whose centres are O, A, B and C. Since ABC is self-conjugate to S,

$$S = a\alpha^2 \cos A + b\beta^2 \cos B + c\gamma^2 \cos C,$$

also since the sides of the triangle of reference are the radical axes of S and U, V, W respectively,

$$U = S - l\alpha I, \quad V = S - m\beta I, \quad W = S - n\gamma I \ldots\ldots(23),$$

whence $\qquad l\alpha - m\beta = 0, \quad m\beta - n\gamma = 0, \quad n\gamma - l\alpha = 0$

are the radical axes of U and V, V and W, W and U respectively. But these are the equations of CF, AD and BE; whence

$$l \sec A = m \sec B = n \sec C = k,$$

and (23) becomes

$$U = S - kI\alpha \cos A \ \&c.$$

The constant k is determined from the fact that the point A is the centre of U, and therefore the pole of the line at infinity; whence $k = 2$ and (23) becomes

$$\left.\begin{aligned} U &= S - 2I\alpha \cos A \\ V &= S - 2I\beta \cos B \\ W &= S - 2I\gamma \cos C \end{aligned}\right\} \ \dots\dots \ \dots\dots\dots\dots(24).$$

215.   *If λ, μ, ν be variable parameters, the circle*

$$\lambda U + \mu V + \nu W = 0$$

cuts the circle S orthogonally.

The radical axes of *any* three circles intersect at the radical centre; and from this point as centre a circle S can be described cutting each of the three circles orthogonally. Also if a fourth circle be described, such that the radical axis of the latter and *any* one of the three circles passes through the radical centre, this circle will be cut orthogonally by S. Hence the circle $\lambda U + \mu V + \nu W = 0$ will be cut orthogonally by S, provided the radical axis of itself and U passes through O. Now

$$\lambda U + \mu V + \nu W = (\lambda + \mu + \nu) S - 2I (\lambda\alpha \cos A + \mu\beta \cos B + \nu\gamma \cos C),$$

and the radical axis of this and U is

$$- (\mu + \nu) \alpha \cos A + \mu\beta \cos B + \nu\gamma \cos C = 0,$$

which obviously passes through the point O, where

$$\alpha \cos A = \beta \cos B = \gamma \cos C.$$

From (24) it follows that the equation of a bicircular quartic may be expressed in the form

$$l (S - 2I\alpha \cos A)^2 + m (S - 2I\beta \cos B)^2$$
$$+ n (S - 2I\gamma \cos C)^2 = 0\dots\dots(25),$$

which shows that the quartic passes through the two circular points at infinity.

The condition that the focal conic (22) should be a parabola is that $l + m + n = 0$, in which case (25) becomes

$$S (l\alpha \cos A + m\beta \cos B + n\gamma \cos C)$$
$$= I (l\alpha^2 \cos^2 A + m\beta^2 \cos^2 B + n\gamma^2 \cos^2 C),$$

which is the equation of a circular cubic.

Centres of Inversion.

216. If the centre and radius δ of the fixed circle whose centre is O be taken as the centre and radius of inversion, it follows from (12) of § 200 that a bicircular quartic is inverted into itself. We shall now show that the vertices A, B, C of the triangle, formed by the intersection of the diagonals of the quadrilateral whose angles are the points where the focal conic intersects the fixed circle, possess the same property; and that the radii of inversion in the three respective cases are the tangents from A, B and C to the fixed circle, that is to say the radii r_1, r_2, r_3 of the circles U, V, W.

Let A be the origin, and AB the axis of x of a Cartesian system of coordinates; then

$$U = x^2 + y^2 - r_1^2,$$
$$V = x^2 + y^2 - 2cx + c^2 - r_2^2,$$
$$W = x^2 + y^2 - 2bx \cos A - 2by \sin A + b^2 - r_3^2.$$

But from § 213

$$c^2 - r_2^2 = 4R^2 \sin C (\sin C - \sin A \cos B)$$
$$= 4R^2 \sin C \sin B \cos A = r_1^2.$$

Similarly $\qquad b^2 - r_3^2 = r_1^2,$

whence if U', V', W' denote the inverses of these circles when the radius of inversion is r_1, we have

$$r^2 U' = - r_1^2 U; \quad r^2 V' = r_1^2 V; \quad r^2 W' = r_1^2 W;$$

which shows that the inverse of the quartic is the same curve as the original quartic.

Focal Conics and Foci.

217. We have shown that four of the foci of a bicircular quartic are the intersections of a circle of inversion with its corresponding focal conic. We shall now prove that the quartic can be generated by taking any one of the three circles U, V or W as the fixed circle, and a conic confocal with the original conic as the focal conic. The intersections of these three circles with their respective focal conics furnish twelve more foci, making altogether sixteen.

In the figure to § 200, let (f', g') be the coordinates of the centre A of the circle U referred to O, and (f_1, g_1) of A referred to E; also let (ξ, η) be current coordinates referred to A, the axes being parallel to those of the focal conic. Then

$$x = \xi + f', \quad y = \eta + g' \atop f_1 = f + f', \quad g_1 = g + g' \right\} \quad \dots\dots\dots\dots(26).$$

Also let

$$P = ff' + gg' + \delta^2 \atop S_1 = \xi^2 + \eta^2 + 2f_1\xi + 2g_1\eta + r_1^2 \right\} \quad \dots\dots\dots(27),$$

where $AO^2 = r_1^2 + \delta^2$. Then (12) can be transformed into

$$(S_1 + 2P)^2 = 4\{a^2(\xi + f')^2 + b^2(\eta + g')^2\},$$

or
$$S_1^2 = 4\{\xi^2(a^2 - P) + \eta^2(b^2 - P) \\ + 2\xi(a^2f' - Pf_1) + 2\eta(b^2g' - Pg_1) \\ + a^2f'^2 + b^2g'^2 - Pr_1^2 - P^2\}\dots\dots\dots\dots(28).$$

The equation

$$\frac{(\xi + f_1)^2}{a^2} + \frac{(\eta + g_1)^2}{b^2} - 1 + \lambda\{(\xi + f')^2 + (\eta + g')^2 - \delta^2\} = 0$$

represents a conic passing through the points P, Q, R, S in which the focal conic intersects the circle of inversion; and by suitably determining λ this conic may be made to represent the two straight lines AP, AQ. Since A is the origin, we must determine λ from the conditions that the coefficients of ξ and η and also the absolute term vanish, which give

$$\frac{f_1}{a^2} + \lambda f' = 0 \quad \dots\dots\dots\dots\dots(29),$$

$$\frac{g_1}{b^2} + \lambda g' = 0 \quad \dots\dots\dots\dots\dots(30),$$

$$\frac{f_1}{a^2} + \frac{g_1}{b^2} - 1 + \lambda(f'^2 + g'^2 - \delta^2) = 0 \dots\dots\dots(31).$$

Multiplying (29) and (30) by f_1, g_1, adding, and taking account of (31) and the first of (27) we obtain

$$\lambda P + 1 = 0\dots\dots\dots\dots\dots(32),$$

whence (29) and (30) become

$$Pf_1 - a^2f' = 0, \quad Pg_1 - b^2g' = 0 \dots\dots\dots(33),$$

which show that the coefficients of ξ and η in (28) vanish.

Again from (33)

$$a^2 f'^2 + b^2 g'^2 = P(f' f_1 + g' g_1)$$

and

$$Pr_1^2 = P(A O^2 - \delta^2)$$
$$= P(f'^2 + g'^2 - \delta^2),$$

whence

$$a^2 f'^2 + b^2 g'^2 - P r_1^2 = P(ff' + gg' + \delta^2)$$
$$= P^2,$$

which shows that the absolute term in (28) vanishes. Hence the equation becomes

$$S_1^2 = 4\{\xi^2(a^2 - P) + \eta^2(b^2 - P)\}.$$

Comparing this with (12), it follows that the quartic can be generated as the envelope of a variable circle which cuts the circle U orthogonally, and whose centre moves on the conic whose equation referred to the centre and axes of the original conic is

$$\frac{x^2}{a^2 - P} + \frac{y^2}{b^2 - P} = 1,$$

which is confocal with the original conic.

218. When any of the points of intersection of one of the four circles of inversion with its respective focal conic are imaginary, the corresponding foci will be imaginary; also if the circle touches its focal conic at one point, the point of contact will be a double focus; and if the circle osculates the conic, the point of contact will be a triple focus. It also follows from § 76 that the sixteen foci cannot all be real; for bicircular quartics, with only two double points, are quartics of the eighth class, and consequently by § 79 possess two real double foci, and four real single ones, which may however unite into one or more multiple foci. Cartesians on the other hand are curves of the sixth class, and therefore by § 80 possess one real triple focus and three real single ones.

219. The form of equation (12) of § 200 shows that the lines drawn through the centre O of the fixed circle which are parallel to the asymptotes of the reciprocal of the focal conic are one pair of double tangents, and that these will be real or imaginary according as the focal conic is a hyperbola or an ellipse, and the results of § 217 show that the remaining three pairs of double tangents are parallel to the asymptotes of the reciprocals of the other three focal conics. The preceding theorem will, however, require modification when the quartic has a third double point.

220. *The foci of the focal conic are the double foci of the quartic.*

In (12) of § 200 write $\beta = x + \iota y$, $\gamma = x - \iota y$, and make the resulting equation homogeneous by multiplying each term by the proper power of α; then the equation

$$\{\beta\gamma + f\alpha(\beta + \gamma) - \iota g\alpha(\beta - \gamma) + \delta^2\alpha^2\}^2 = \alpha^2\{a^2(\beta + \gamma)^2 - b^2(\beta - \gamma)^2\}$$

represents a quartic having a pair of imaginary nodes at the points B and C of the triangle of reference.

The nodal tangents at B and C are

$$(\gamma + f\alpha - \iota g\alpha)^2 = (a^2 - b^2)\alpha^2$$

and

$$(\beta + f\alpha + \iota g\alpha)^2 = (a^2 - b^2)\alpha^2.$$

Retransforming to Cartesian coordinates and putting $\alpha = 1$, it follows that the nodal tangents at the two circular points are

$$x + f \pm ae - \iota(y + g) = 0,$$
$$x + f \pm ae + \iota(y + g) = 0,$$

which intersect at the two real points

$$x = ae - f, \qquad y = -g,$$
$$x = -ae - f, \quad y = -g.$$

These are the coordinates referred to O of the foci of the focal conic, and therefore, by § 79, these points are the two double foci of the quartic.

Putting $e = 0$, it follows from § 200 and the preceding paragraph that when the focal conic reduces to a circle, the centre of the latter is the triple focus of a cartesian.

221. *If r_1, r_2, r_3 be the distances of any point on a bicircular quartic from any three real foci, then*

$$lr_1 + mr_2 + nr_3 = 0.$$

Let the point r_1 be taken as the origin, and let the axis of x pass through r_2; then

$$r_1^2 = r^2,$$
$$r_2^2 = r^2 - 2ax + a^2,$$
$$r_3^2 = r^2 - 2(ex + fy) + e^2 + f^2,$$

and therefore the required locus is

$$\{(l^2 + m^2 - n^2)\, r^2 - 2m^2 ax + 2n^2\,(ex + fy) + m^2 a^2 - n^2\,(e^2 + f^2)\}^2$$

$$= 4l^2 m^2 r^2 \,(r^2 - 2ax + a^2)^2 \quad\dots\dots\dots\dots\dots(34),$$

which is the equation of a bicircular quartic. To prove that the three points are foci, we shall show that the line $x + \iota y = 0$ is a tangent to the quartic. Substituting ιx for y in (34) it becomes

$$\{2n^2\,(e + \iota f)\, x - 2m^2 ax + m^2 a^2 - n^2\,(e^2 + f^2)\}^2 = 0,$$

which is a perfect square, and therefore shows that the line $x + \iota y = 0$ touches the curve at the imaginary point determined by this equation.

222. It follows from §§ 202 and 81 that if a conic be inverted from any point O, the point O and the two inverse points of the foci of the conic are foci of the quartic. We shall now prove geometrically that if P be any point on the quartic OP, SP and HP are connected by a linear relation.

Let C', S', H' denote the centre and foci of the conic, P' any point on the conic, $2a$ its major axis; then

$$\frac{H'P'}{HP} = \frac{OP'}{OH}; \quad \frac{S'P'}{SP} = \frac{OP'}{OS};$$

also since $S'P' + H'P' = 2a$, we obtain

$$\frac{2a}{OP'} = \frac{HP}{OH} + \frac{SP}{OS} \quad\dots\dots\dots\dots\dots(35),$$

or if k is the radius of inversion

$$2aOP/k^2 = HP/OH + SP/OS \quad\dots\dots\dots\dots(36),$$

which is the required linear relation.

Let O coincide with the centre C of the conic; then

$$OH' = OS' = ae,$$

whence (35) becomes

$$\tfrac{1}{2}e\,(SP + HP) = CP.$$

When the conic is a hyperbola, this becomes

$$\tfrac{1}{2}e\,(SP - HP) = CP,$$

also $\qquad\qquad SP^2 + HP^2 = 2\,(CP^2 + CS^2),$

whence $\qquad\qquad SP\,.\,HP = CP^2\,(1 - 2/e^2) + CS^2.$

When the conic is a rectangular hyperbola, $e = \sqrt{2}$, and we obtain

$$SP \cdot HP = CS^2,$$

which is the well known focal property of the lemniscate.

Circular Cubics.

223. We have shown in § 200 that when the focal conic is a parabola, a bicircular quartic becomes a circular cubic; and also that the inverse of a bicircular quartic from any point on the curve is a circular cubic. Conversely the inverse of a circular cubic from any point not on the curve is a bicircular quartic. When the centre of inversion is on the cubic, the curve inverts into another circular cubic unless the point is a double point, in which case the cubic inverts into a conic which is a hyperbola, a parabola or an ellipse according as the double point is a node, a cusp or a conjugate point. All the above results follow from the general equation (4) of § 198 of a bicircular quartic.

The equation of a circular cubic in Cartesian coordinates is

$$x\,(r^2 + 2fx + 2gy + \delta^2) + 2ar^2 = 0 \quad \dots\dots\dots\dots(1),$$

the origin being the centre O of the fixed circle; $(f,\ g)$ are the coordinates of O referred to the focus S of the focal parabola, and a is the focal distance of the latter. If (1) be transformed to polar coordinates we obtain

$$r^2 + 2r\,(f\cos\theta + g\sin\theta + a\sec\theta) + \delta^2 = 0\dots\dots\dots(2),$$

from which it follows that if

$$f\cos\alpha + g\sin\alpha + a\sec\alpha + \delta = 0 \dots\dots\dots\dots\dots(3),$$

the line $\theta = \alpha$ intersects the curve in two coincident points. If we now transfer the origin to the point $\delta\cos\alpha,\ \delta\sin\alpha$, where δ is given by (3), the linear terms will not vanish; whence the new origin is not a double point, but is the point of contact of one of the tangents from O to the curve. Accordingly the cubic cannot have a double point unless $\delta = 0$; in which case the equation of the tangents at the double point is

$$(a + f)\,x^2 + gxy + ay^2 = 0,$$

and the double point will be a node, a cusp or a conjugate point according as

$$g^2 > \text{ or } = \text{ or } < 4a(a+f),$$

that is according as the point O is without, upon or within the focal parabola. In this case the cubic will be the inverse of a conic with respect to a point on the curve.

All circular cubics have only one real asymptote, viz. the line

$$x + 2a = 0.$$

From (2) we obtain

$$r_1 r_2 = \delta^2 \dots\dots\dots\dots\dots\dots\dots\dots\dots(4),$$

$$\tfrac{1}{2}(r_1 + r_2) = -(f\cos\theta + g\sin\theta + a\sec\theta) \dots\dots\dots(5).$$

Equation (4) shows that if OPQ be any chord, the rectangle $OP \cdot OQ = \delta^2$; also that the lengths of the tangents drawn from O to the curve are all equal to the radius of the fixed circle. Equation (5) shows that the locus of the middle point of PQ is the circular cubic

$$r^2 x + (a+f)x^2 + gxy + ay^2 = 0,$$

whose node and nodal tangents are identical with those of the first cubic when $\delta = 0$.

224. We have shown in § 121 that one of the forms of the equation of a circular cubic in trilinear coordinates is

$$u_1 S = I u_2 \dots\dots\dots\dots\dots\dots\dots\dots(6),$$

where u_n is a ternary quantic of degree n.

The form of (6) shows (i) that the cubic passes through the two circular points at infinity; (ii) that it passes through the point where the line u_1 intersects the line at infinity, from which it follows that the line $u_1 = 0$ is parallel to the asymptote; (iii) that the cubic passes through the points of intersection of the conic $u_2 = 0$ with the circle $S = 0$ and the line $u_1 = 0$. It also follows that a circle cannot intersect a circular cubic at more than four points which are at a finite distance from one another.

225. The following proposition is of fundamental importance in the theory of circular cubics.

If a circle intersect a circular cubic in four points A, B, C, D, the three straight lines which respectively join the points, where the three pairs of straight lines AB, CD; BC, AD; CA, BD again intersect the cubic, are parallel to the asymptote.

If S be taken as the circle circumscribing ABC, the conic u_2 must also circumscribe this triangle, whence

$$u_2 = l\beta\gamma + m\gamma\alpha + n\alpha\beta,$$

$$u_1 = \lambda\alpha + \mu\beta + \nu\gamma.$$

Let the lines AB, CD meet the cubic in E and F; then since D is the fourth point in which S and u_2 intersect, the equation of CD is

$$c\,(l\beta + m\alpha) = n\,(a\beta + b\alpha) \quad\dots\dots\dots\dots\dots(7).$$

To find the third point F where CD intersects the cubic, substitute the left-hand side of (7) in the term u_2 in the cubic and it reduces to

$$(cu_1 - nI)\,S = 0,$$

which shows that the line

$$cu_1 - nI = 0 \quad\dots\dots\dots\dots\dots\dots\dots(8)$$

passes through F. Putting $\gamma = 0$ in (8) and also in (6) it follows that (8) passes through E; whence (8) is the equation of EF. The form of (8) shows that EF is parallel to the asymptote.

226. If A and B coincide, AE is the tangent at A, whence :—

If a circle touch the cubic at A and intersect it at C and D, the tangent at A and the line CD intersect the cubic at two points E and F, such that EF is parallel to the asymptote.

227. Let C and D as well as A and B coincide, then :—

If a circle touch the cubic at two points, the line joining the two points, where the tangents at the points of contact cut the cubic, is parallel to the asymptote.

228. Let A, B and C coincide, then :—

If the chord of curvature intersect the cubic in F, and the tangent to the cubic and its circle of curvature meet the curve in E, the line EF is parallel to the asymptote.

229. *If a straight line parallel to the asymptote of a circular cubic cut the curve in a and c, and if the tangents at a and c cut the curve in A and C, then the four points $AacC$ lie on a circle; also AC intersects the cubic at the point where it is cut by its asymptote.*

Let A, B, C, D be the points where any circle intersects the
cubic; let AB, CD intersect the cubic
in a and c; and AC, BD in R and R'.
Let B move up to coincidence with a,
and D with c. Then Aa, Cc are the
tangents at a and c, and the four
points $AacC$ lie on a circle.

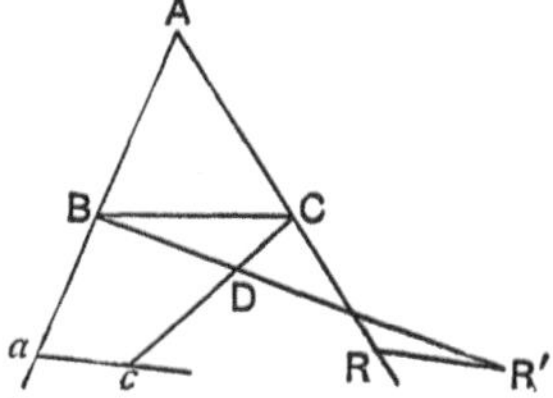

When B and D coincide with a and c, BD coincides with ac;
but since a line which is parallel to the asymptote cannot cut the
curve in more than two points at a finite distance from one
another, the point R' must move off to infinity. Hence the line
RR', which by § 225 is parallel to the asymptote, cuts the cubic
in only one finite point R, and therefore it must be the asymptote.

230. Let the points A and C coincide; then :—

*If a tangent be drawn to the cubic from the point where it is
cut by its asymptote, and if from the point of contact A two tangents
be drawn to the cubic touching it in a and c, the circle circumscribing Aac will touch the cubic in A, and the line ac will be parallel
to the asymptote.*

231. *If a circular cubic Σ be inverted from any point O on
itself into a circular cubic Σ', the osculating circle of Σ at O will
invert into the asymptote of Σ', and vice versâ.*

The osculating circle intersects the cubic in three coincident
points at O, and one finite point P; whence the circle inverts
into a line cutting the inverse cubic in one finite point P' and
touching it at two coincident points at infinity; whence the
inverse of the osculating circle is the asymptote of Σ'.

232. *If the cubic be inverted from the point O where the
asymptote cuts the curve, the point O will be a point of inflexion on
the inverse curve.*

It follows from (6) that the equation

$$(ax + by)\, S + \lambda\, (ax + by)^2 + ex + fy = 0$$

represents a circular cubic whose asymptote is the line

$$ax + by = 0.$$

The inverse cubic is

$$(ax + by)\, S' + \lambda k^2 (ax + by)^2 + r^2 (ex + fy) = 0,$$

where S' is the inverse of the circle S; from which it follows that the line $ax + by = 0$ has a contact of the second order with the inverse cubic at the origin, and the latter is therefore a point of inflexion.

233. *If three tangents be drawn to a circular cubic from the point O in which the cubic cuts its asymptote, the three points of contact will lie on a circle which passes through O.*

We have shown in § 92 that from a point of inflexion O three tangents can be drawn to a cubic, and that the three points of contact lie on a straight line. Hence inverting with respect to O, the theorem at once follows from § 232.

234. *Every circular cubic passes through the four centres of inversion, and also through the feet of the perpendiculars of the triangle formed by joining any three centres of inversion.*

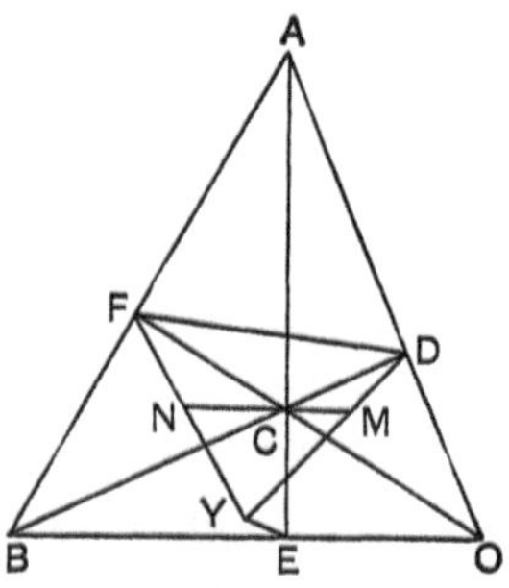

We have shown in § 213 that if A, B, C, O be the four centres of inversion, any one of these points is the orthocentre of the triangle formed by joining the other three. Also the equation of the cubic referred to ABC is

$$S\left(l\alpha \cos A + m\beta \cos B + n\gamma \cos C\right)$$

$$= I\left(l\alpha^2 \cos^2 A + m\beta^2 \cos^2 B + n\gamma^2 \cos^2 C\right)\ldots\ldots(9),$$

where
$$S = a\alpha^2 \cos A + b\beta^2 \cos B + c\gamma^2 \cos C\ldots\ldots\ldots\ldots(10),$$

and
$$l + m + n = 0\ldots\ldots\ldots\ldots\ldots\ldots\ldots(11).$$

Putting $\beta = \gamma = 0$ in (9), it follows that (9) vanishes by virtue of (10); whence the cubic passes through A; similarly it passes through B and C. Also since the coordinates of O are proportional to $\sec A$, $\sec B$, $\sec C$ it follows that the cubic passes through O.

To find where AC cuts the cubic, put $\beta = 0$, and (9) reduces to

$$\alpha\gamma \, (lc \cos A - na \cos C)(\alpha \cos A - \gamma \cos C) = 0,$$

the last factor of which is the equation of BE.

235. *The tangents to the cubic at the four centres of inversion are parallel to the asymptote.*

Since the four points B, E, C, F lie on the cubic and also on a circle, it follows from § 225 that the line joining the third points in which BF and EC intersect the cubic is parallel to the asymptote; but since these lines intersect on the cubic at A, the tangent at A is parallel to the asymptote.

Since the tangent at A is the coefficient of α^2 in the equation of the cubic, it follows from (9) and (11) that its equation is

$$\beta \, (m \sin C + n \cos A \sin B) + \gamma \, (n \sin B + m \cos A \sin C) = 0$$
$$\dots\dots\dots\dots(12).$$

A direct proof may be given as follows. The form of (9) shows that the line

$$l\alpha \cos A + m\beta \cos B + n\gamma \cos C = 0\dots\dots\dots\dots(13),$$

or $u = 0$, is parallel to the asymptote. The equation of any line parallel to (13) is $u + kI = 0$; and if we determine k so that this line passes through A, we shall obtain (12), which is the tangent at A.

236. *The tangents at D, E, F intersect at a point Y on the nine-point circle, which is common to the four triangles formed by joining the centres of inversion.*

Let the tangents at D and F intersect in Y; join EY; also let the tangent at C cut YD, YF in M and N. Then since D and F are the inverse points of C with respect to B and O,

$$YDC = MCD; \quad YFC = NCF = OCM,$$

whence $\qquad\qquad YDC + YFC = OCD = \tfrac{1}{2}\pi - B\dots\dots\dots\dots(14)$.

Also $\qquad\qquad \pi - DYF = YDC + YFC + CDF + CFD\dots\dots\dots(15)$.

But $\qquad\qquad CDF = A; \quad CFD = CAD = C - \tfrac{1}{2}\pi\dots\dots\dots\dots(16)$.

Substituting from (14) and (16) in (15) we get

$$DYF = 2B.$$

But from the geometry of the nine-point circle, it is known that this circle passes through DEF, and that $DEF = 2B$; whence Y is a point on the nine-point circle.

Whence also $EYD = EFD = 2C$, and therefore YE is the tangent at E.

237. *If the sides DE, EF, FD of the triangle DEF be produced to meet the cubic in F', D', E', the lines DD', EE', FF' are parallel to the asymptote.*

Since the four points A, F, E, O lie on the cubic and also on a circle, and AO and FE cut the cubic again in D and D', it follows from § 225 that DD' is parallel to the asymptote.

238. *The point of intersection Y of the tangents at D, E, F is the point where the cubic is cut by its asymptote.*

Let the nine-point circle cut the cubic in a fourth point H, and let HD cut the cubic again in D''. Then by § 225 $D'D''$ is parallel to the asymptote; but by § 237 $D'D$ is also parallel to the asymptote, whence D'' coincides with D, and HD is the tangent at D. Accordingly H must coincide with Y. This shows that Y is also a point on the cubic.

Also since the points of contact D, E, F of the tangents from Y lie on a circle passing through Y, it follows from § 233 that Y is the point where the asymptote cuts the cubic. Hence :—

The nine-point circle, common to the triangles formed by joining any three of the centres of inversion, passes through the point where the cubic is cut by its asymptote.

The preceding theorems furnish the following construction for finding the four centres of inversion. From the point where the cubic cuts its asymptote, draw three tangents and let D, E, F be their points of contact; then the centres of the inscribed and the three escribed circles of the triangle DEF are the four centres of inversion.

239. *The common nine-point circle of the triangle formed by joining any three of the four centres of inversion passes through the focus of the focal parabola, that is through the double focus of the cubic.*

Since by § 216 the triangle is self-conjugate to the parabola, the equation of the latter may be written

$$l\alpha^2 + m\beta^2 + n\gamma^2 = 0 \quad\ldots\ldots\ldots\ldots\ldots\ldots(17),$$

where

$$a^2/l + b^2/m + c^2/n = 0 \quad\ldots\ldots\ldots\ldots\ldots(18).$$

Let D_1, E_1 and F_1 be the middle points of BC, CA, AB; then the equation of D_1E_1 is

$$a\alpha + b\beta - c\gamma = 0 \quad\ldots\ldots\ldots\ldots\ldots\ldots(19),$$

and (18) is the condition that (19) should touch (17). Hence the parabola touches the sides of the triangle $D_1E_1F_1$, and therefore its focus lies on the circle circumscribing $D_1E_1F_1$, that is upon the nine-point circle of ABC.

240. *The directrices of the four focal parabolas pass respectively through the centres of the four circles circumscribing the four triangles formed by joining the centres of inversion.*

The equation of the directrix of (17) is*

$$l(m+n)\,\alpha/a + m(n+l)\,\beta/b + n(l+m)\,\gamma/c = 0.$$

The condition that this line should pass through the centre of the circle circumscribing ABC, whose coordinates are proportional to $\cos A$, $\cos B$ and $\cos C$, is

$$lm(\cot A + \cot B) + mn(\cot B + \cot C) + nl(\cot C + \cot A) = 0,$$

which is the same thing as (18), which is the condition that (17) should be a parabola.

On the Points of Inflexion.

241. The general equation of a circular cubic may be written in the form

$$Su_1 = Iu_2 \quad\ldots\ldots\ldots\ldots\ldots\ldots\ldots(20),$$

where S is the circumscribing circle, and

$$u_1 = \lambda\alpha + \mu\beta + \nu\gamma \quad\ldots\ldots\ldots\ldots\ldots\ldots\ldots\ldots(21),$$

$$u_2 = l\alpha^2 + m\beta^2 + n\gamma^2 + 2l'\beta\gamma + 2m'\gamma\alpha + 2n'\alpha\beta \quad\ldots\ldots(22).$$

* Ferrers' *Trilinear Coordinates*, p. 93.

Let B and C be two real points of inflexion, and let AB, AC be the tangents at B and C. Then if $\beta = 0$, the cubic must reduce to $\alpha^3 = 0$, which requires that

$$n = 0, \quad 2m' = \nu b/c,$$

$$lc = \lambda b - 2m'a = b\,(\lambda c - \nu a)/c \dots\dots\dots\dots(23).$$

The conditions that the cubic should reduce to $\alpha^3 = 0$ when $\gamma = 0$ are that

$$m = 0, \quad 2n' = \nu c/b,$$

$$lb = c\,(\lambda b - \mu a)/b \ \dots\dots\dots\dots\dots(24).$$

The third real point of inflexion D must lie on BC, whence putting $\alpha = 0$ in (20), the equation of AD is

$$(\mu\beta + \nu\gamma)\,a = 2l'\,(b\beta + c\gamma).$$

If D is at infinity, AD must be parallel to BC, whence

$$\mu/b = \nu/c = k \ \dots\dots\dots\dots\dots\dots(25).$$

Using this in (23) and (24) we get

$$(\lambda - ka)\,(b^2 - c^2) = 0\dots\dots\dots\dots\dots(26).$$

The solution $\lambda = ka$ must be rejected, because it leads to the cubic breaking up into a conic and the line at infinity ; the other solution shows that $b = c$, whence $\mu = \nu$, and the equation of u_1 becomes

$$\lambda\alpha + \mu\,(\beta + \gamma) = 0,$$

which is parallel to BC. These results show that when a circular cubic has three real points of inflexion, one of which is at infinity :—

(i) *The tangents at the two other points of inflexion, together with their chord of contact, form an isosceles triangle of which the chord of contact is the base.*

(ii) *The line joining the points of inflexion is parallel to the asymptote.*

242. The following is an example of tangential coordinates.

If through any point O on a circular cubic a line be drawn cutting the cubic in P and Q, and RX be drawn perpendicular to PQ through the middle point R of PQ, the envelope of RX is a parabola.

Taking O as the origin, the equation of the cubic is

$$r^2 v_1 + u_2 + u_1 = 0,$$

which in polar coordinates becomes

$$r^2 (E \cos \theta + F \sin \theta) + 2r (A \cos^2 \theta + B \sin 2\theta + C \sin^2 \theta)$$
$$+ G \cos \theta + H \sin \theta = 0,$$

whence

$$OR = \tfrac{1}{2}(r_1 + r_2) = -\frac{A \cos^2 \theta + B \sin 2\theta + C \sin^2 \theta}{E \cos \theta + F \sin \theta}.$$

Let RX meet the axes in X and Y; and let $OR = \rho$, then

$$\rho \xi = \cos \theta, \quad \rho \eta = \sin \theta,$$

whence $\qquad A\xi^2 + 2B\xi\eta + C\eta^2 + E\xi + F\eta = 0,$

which is the tangential equation of a parabola.

CHAPTER X.

SPECIAL QUARTICS.

243. HAVING discussed the general theory of quartic curves and also that of bicircular quartics, we shall proceed to consider the properties of a variety of well known curves of this degree. It will further be shown in Chapter XII. that all the projective properties of these curves may be generalized by projection; and in particular that the theory of all quartics having three biflecnodes, a node and a pair of cusps, or three cusps may be deduced from the properties of the lemniscate, the limaçon and the cardioid respectively.

The Cassinian.

244. The Cassinian, or the oval of Cassini as it is some-
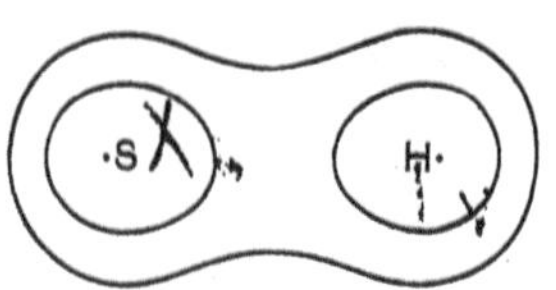
times called, is the locus of a point which moves so that the product of its distances from two fixed points is constant. The two fixed points are, as will hereafter be shown, *triple foci.*

To find the equation of the Cassinian.

Let S and H be the foci, O the middle point of SH; let $SH = 2c$, and let a be a constant such that

$$SP \cdot HP = a^2 - c^2 \dots\dots\dots\dots\dots(1),$$

then if (x, y) be the coordinates of P referred to O,

$$SP^2 = (x - c)^2 + y^2, \quad HP^2 = (x + c)^2 + y^2,$$

whence

$$(x^2 + y^2 + c^2)^2 - 4c^2 x^2 = (a^2 - c^2)^2 \dots\dots\dots\dots(2).$$

Comparing this with (4) of § 198, it follows that the Cassinian is a bicircular quartic. Equation (2) may also be written

$$r^4 + 2c^2 (a^2 - r^2 \cos 2\theta) - a^4 = 0 \quad \ldots\ldots\ldots\ldots\ldots(3).$$

245. The Cassinian is also included amongst the curves given by the equation

$$\xi + \iota\eta = \tfrac{1}{2} \log \{(x + \iota y)^2/c^2 - 1\}$$

which is equivalent to the two equations

$$(x^2 + y^2)^2 - 2c^2 (x^2 - y^2) + c^4 (1 - \epsilon^{4\xi}) = 0,$$

and

$$x^2 - y^2 - c^2 = 2xy \coth 2\eta,$$

the first of which represents a family of confocal Cassinians, and the second a family of rectangular hyperbolas which pass through the foci of the former and cut them orthogonally.

246. The Cassinian always cuts the axis of x in the two real points $x = \pm a$; and will also cut it in two other points which will be real or imaginary according as $c\sqrt{2} >$ or $< a$. The curve also cuts the axis of y in four imaginary points or in two real and two imaginary ones according as $c\sqrt{2} >$ or $< a$. Accordingly when $c\sqrt{2} > a$, the Cassinian is an exodromic curve consisting of two detached ovals, each of which encloses one of the foci; but when $c\sqrt{2} < a$, the curve is unipartite and perigraphic; and the internal and external curves in the figure to § 244 show the forms of the Cassinian in the two respective cases. When $c\sqrt{2} = a$, the curve becomes the lemniscate of Bernoulli, and the origin is a real biflecnode. The form of the curve is shown in § 253.

Transforming (2) into trilinear coordinates by taking the lines $x \pm \iota y = 0$ as two of the sides of the triangle of reference and the line at infinity as the third side, the equation becomes

$$(\beta\gamma + c^2 I^2)^2 - c^2 I^2 (\beta + \gamma)^2 = (a^2 - c^2)^2 I^4,$$

from which it follows that the Cassinian is a binodal quartic having a pair of biflecnodes at the circular points. Hence Plücker's numbers are $n = 4$, $m = 8$, $\delta = 2$, $\kappa = 0$, $\tau = 8$, $\iota = 12$; but since four of the points of inflexion are situated at the biflecnodes, the curve has only eight independent points of inflexion; and it will be shown in § 251 that four of these must be imaginary, whilst the remaining four may be all real or all imaginary, or may coalesce into two real points of undulation.

247. The nodal tangents at the circular points are

$$\gamma^2 = c^2 I^2, \quad \beta^2 = c^2 I^2 \,;$$

or, in Cartesian coordinates,

$$(x - \iota y)^2 = c^2, \quad (x + \iota y)^2 = c^2,$$

which intersect at the points $x = \pm c$, $y = 0$; also, since both tangents are stationary tangents, their points of intersection are triple foci.

Since every binodal quartic must have eight real foci, of which two or more may unite into multiple foci, it follows that the Cassinian must have two single foci. Their positions may be found by determining the conditions that the line $x - \alpha \pm \iota (y - \beta) = 0$ should be a tangent to the curve.

Writing $p = \alpha + \iota \beta$, and eliminating y from (2), we shall obtain

$$(p^2 - c^2)(4a^2 - 4px + p^2 - c^2) = (a^2 - c^2)^2,$$

which will have equal roots if

$$(p^2 - c^2)(p^2 c^2 + a^4 - 2a^2 c^2) = 0 \,;$$

the first factor gives $\alpha = \pm c$, $\beta = 0$, which are the coordinates of the triple foci S and H. As regards the other factor, we observe that when $c \sqrt{2} > a$, in which case the Cassinian is bipartite, we obtain $pc = \pm a (2c^2 - a^2)^{\frac{1}{2}}$, which gives $\alpha = \pm a (2c^2 - a^2)^{\frac{1}{2}}/c$, $\beta = 0$; hence in this case there is a pair of real single foci on the axis of x. But when $c \sqrt{2} < a$, in which case the Cassinian is unipartite, we obtain $pc = \pm \iota a (a^2 - 2c^2)^{\frac{1}{2}}$, which gives $\alpha = 0$, $\beta = \pm a (a^2 - 2c^2)^{\frac{1}{2}}/c$; hence in this case there is a pair of real foci on the axis of y. When $a = c \sqrt{2}$, the curve becomes a lemniscate, and the origin is a double focus formed by the union of the foregoing pair of single foci. It can also be shown that the two single foci lie inside or outside the curve, according as the Cassinian is bipartite or unipartite.

We shall now give some properties of the curve.

248. *If P be a point on the curve, straight lines drawn from the foci perpendicular to SP, HP will meet the tangent at P in points equidistant from P.*

Let $SP = r$, $HP = r'$; draw SK, HK' perpendicular to SP, HP meeting the tangent at P in K, K'. Let $SPK = \phi$, $HPK' = \phi'$.

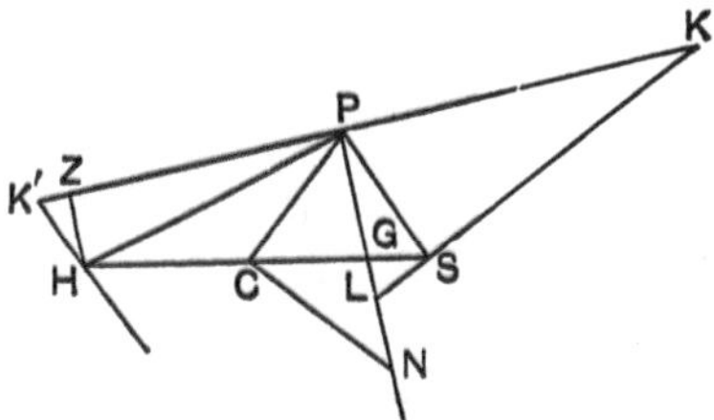

Then $rr' = a^2 - c^2$; and since $dr/ds = \cos\phi$, $-dr'/ds = \cos\phi'$, we obtain $r \sec\phi = r' \sec\phi'$; whence $PK = PK'$.

249. *If the normal at P meet SH in G, and C be the middle point of SH, the angles SPG and CPH are equal; also*

$$SP^2 : HP^2 :: SG : HG.$$

Draw PC meeting SH in C, such that the angle $CPH = SPG$; then

$$\frac{HC}{r'} = \frac{\sin CPH}{\sin C} = \frac{\sin SPG}{\sin C} = \frac{\cos\phi}{\sin C},$$

whence
$$HC \sin C = r' \cos\phi.$$

Similarly $SC \sin C = r \cos\phi' = HC \sin C$, since $r' \cos\phi = r \cos\phi'$; whence $SC = HC$. Therefore C is the middle point of SH.

Again, if SY, HZ are perpendicular to PK,

$$SP^2 = PK \cdot PY; \quad HP^2 = PK' \cdot PZ,$$

and $PK = PK'$ whence

$$SP^2 : HP^2 :: PY : PZ,$$
$$:: SG : HG.$$

250. *Straight lines are drawn from S, H and C perpendicular to SP, HP, CP respectively, meeting the normal at P in L, M and N; prove that if ρ be the radius of curvature at P,*

$$\frac{1}{PL} + \frac{1}{PM} = \frac{1}{PN} + \frac{1}{\rho}.$$

Let $PSH = \theta$, $PHS = \theta'$, $PCS = \chi$, $PGS = \psi$; then

$$PL = r \operatorname{cosec}\phi = -\frac{ds}{d\theta},$$

$$PM = \frac{ds}{d\theta'}, \quad PN = \frac{ds}{d\chi},$$

whence
$$\frac{1}{PL} + \frac{1}{PM} = -\frac{d\theta}{ds} + \frac{d\theta'}{ds}.$$

Also
$$\chi = \theta' + CPH = \theta' + SPG = \theta' + \tfrac{1}{2}\pi - \phi,$$
$$\psi = \theta' + \tfrac{1}{2}\pi - \phi',$$
$$\theta + \theta' = \phi + \phi',$$

whence
$$\chi + \psi = \theta' - \theta + \pi,$$

and therefore
$$\frac{d\theta'}{ds} - \frac{d\theta}{ds} = \frac{d\chi}{ds} + \frac{d\psi}{ds},$$

or
$$\frac{1}{PL} + \frac{1}{PM} = \frac{1}{PN} + \frac{1}{\rho}.$$

251.　*The equation of the curve in terms of p and r is*

$$rp\,(a^2 - c^2) + a^2 c^2 = \tfrac{1}{2}(r^4 + a^4).$$

Let $CP = r$, p the perpendicular from C on to the tangent at P. Draw SB, HB' perpendicular to CP; then

$$\frac{1}{PL} + \frac{1}{PM} = \frac{\sin\phi}{SP} + \frac{\sin\phi'}{HP}$$

$$= \frac{HP\sin\phi + SP\sin\phi'}{a^2 - c^2}$$

$$= \frac{PB' + PB}{a^2 - c^2}$$

$$= \frac{2r}{a^2 - c^2},$$

and
$$\frac{1}{PN} = \frac{p}{r^2},$$

whence, by § 250,
$$\frac{2r}{a^2 - c^2} = \frac{p}{r^2} + \frac{1}{r}\frac{dp}{dr}.$$

Integrating
$$rp = \frac{r^4}{2\,(a^2 - c^2)} + A.$$

To find the constant, we observe that at each of the vertices r and p are each equal to a, whence $A = \tfrac{1}{2}(a^4 - 2a^2 c^2)/(a^2 - c^2)$ and

$$rp\,(a^2 - c^2) + a^2 c^2 = \tfrac{1}{2}(r^4 + a^4).$$

This shows that the radius of curvature may be expressed in the form

$$\rho = \frac{2\,(a^2 - c^2)\,r^3}{3r^4 - a^4 + 2a^2 c^2}.$$

At a point of inflexion $\rho = \infty$, whence the eight points of inflexion are given by the equation $3r^4 = a^4 - 2a^2c^2$, which in combination with (3) gives

$$\cos 2\theta = \pm\, a\, (a^2 - 2c^2)^{\frac{1}{2}}/c^2\, \sqrt{3}.$$

The positive sign may give a real value of θ, but the corresponding value of r will be imaginary. Taking the lower sign, there will be four real points of inflexion provided the value of $\cos 2\theta$ is less than unity, which requires that $c\sqrt{3} > a$. Hence the conditions for four real points of inflexion are that $c\sqrt{3} > a > c\sqrt{2}$. When $c\sqrt{3} = a$, the Cassinian has two real points of undulation.

The preceding argument shows that it is possible for a quartic to have all its points of inflexion imaginary; since an anautotomic quartic which is approximating to the form of a Cassinian having eight imaginary points of inflexion, must have all these singularities imaginary.

252. *A Cassinian is described whose foci are the points of intersection of the directrix with the asymptotes of a hyperbola; prove that the tangents at the points where it meets the auxiliary circle are tangents to the hyperbola, and that the normals at these points pass through the focus of the hyperbola.*

Let K, K' be the foci of the Cassinian, C the centre of the hyperbola, CX its transverse axis, P the point where the Cas-

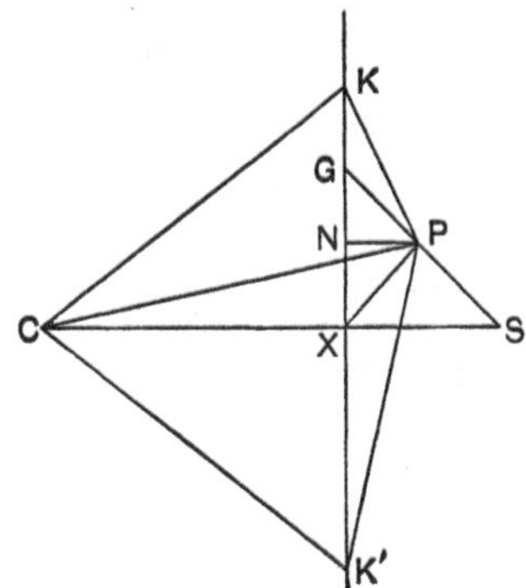

sinian meets the auxiliary circle. Let GPS be the normal at P; $XCK = \alpha$, $XCP = \beta$; draw PN perpendicular to KK'; join KP, $K'P$, CP.

By § 249

$$KPG = K'PX,$$

also
$$CPK = CKP = \tfrac{1}{2}\pi - \tfrac{1}{2}(\alpha - \beta)$$
$$= \tfrac{1}{2}\pi - KK'P = K'PN,$$

whence
$$KPN = K'PC,$$

and therefore
$$CSP = GPN = XPC,$$

whence the triangles CSP and CPX are similar, and therefore $CS \cdot CX = CP^2 = CK^2$. Hence S is the focus of the hyperbola; also the tangent at P to the Cassinian, which is perpendicular to SP, touches the hyperbola.

The Lemniscate of Bernoulli.

253. When $a = c\sqrt{2}$, the Cassinian becomes the lemniscate of Bernoulli, and its equation is

$$r^2 = a^2 \cos 2\theta,$$

or
$$(x^2 + y^2)^2 = a^2(x^2 - y^2).$$

The form of the curve is that of a figure of eight, the origin being a biflecnode, the tangents at which are at right angles.

The lemniscate also possesses the double property of being the inverse and also the pedal of a rectangular hyperbola with respect to its centre.

All the properties which we have already proved for the Cassinian hold good in the case of the lemniscate; we add the following additional ones, which the reader can easily prove.

Let P be the foot of the perpendicular from the centre C on to the tangent at any point Q of a rectangular hyperbola; CY the

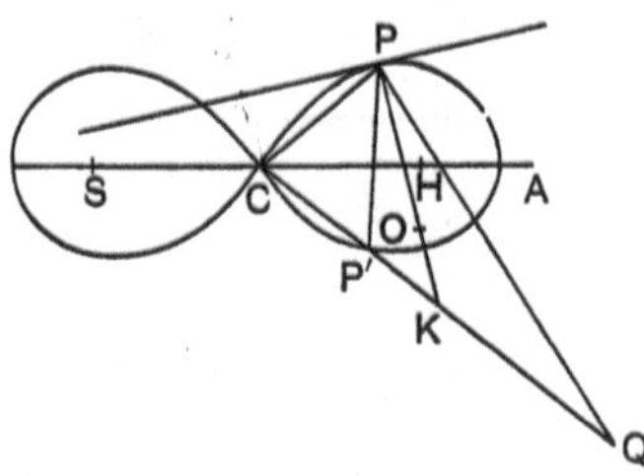

perpendicular from C on to the tangent at P to the lemniscate; also let

$$\phi = CPY, \quad \theta = PCA, \quad \chi = YCA;$$
$$p = CY, \quad CA = a,$$

then
$$QCA = ACP = \theta,$$

$$\phi = CPY = CQP = \tfrac{1}{2}\pi - 2\theta \quad \dots\dots\dots\dots\dots(1),$$

$$\chi = \theta + \tfrac{1}{2}\pi - \phi = 3\theta \quad \dots\dots\dots\dots\dots\dots(2),$$

$$r^3 = a^2 p \quad \dots\dots\dots\dots\dots\dots\dots\dots(3),$$

$$p^{\frac{2}{3}} = a^{\frac{2}{3}} \cos \tfrac{2}{3}\chi \quad \dots\dots\dots\dots\dots\dots(4).$$

Equation (4) is the pedal and also the tangential polar equation of the lemniscate.

Also since the p and r equation of a rectangular hyperbola is $CQ \cdot CP = a^2$, it follows from (3) that

$$\rho = r\,\frac{dr}{dp} = \frac{a^2}{3r} = \tfrac{1}{3}CQ \quad \dots\dots\dots\dots\dots(5).$$

The reciprocal polar is the curve

$$c^{\frac{2}{3}} = r^{\frac{2}{3}} \cos \tfrac{2}{3}\theta \dots\dots\dots\dots\dots\dots(6),$$

and the tangential equation is

$$27 a^4 (\xi^2 + \eta^2)^2 = \{4 - a^2 (\xi^2 - \eta^2)\}^3 \dots\dots\dots\dots(7),$$

which shows that the curve is of the sixth class, a result which follows from the fact that it belongs to species VII.

254. *To find the p and r equation referred to a focus.*

Let $SP = r$, $HP = r'$, $CP = R$; then since C is the middle point of SH,

$$r^2 + r'^2 = 2R^2 + a^2,$$

also
$$2rr' = a^2,$$

whence if $r' > r$,

$$R\sqrt{2} = r' - r = \tfrac{1}{2}a^2/r - r.$$

By (5)
$$\frac{1}{r}\frac{dp}{dr} = \frac{1}{\rho} = \frac{3R}{a^2} = \frac{3}{2r\sqrt{2}} - \frac{3r}{a^2\sqrt{2}};$$

whence
$$2\sqrt{2}a^2 p = (3a^2 - 2r^2)\, r.$$

255. *The angular points of an equilateral triangle move round the circumference of a circle; prove that the locus of the foci of all rectangular hyperbolas which circumscribe the triangle and have a given centre is a lemniscate.*

Let ABC be the triangle, N its orthocentre, O the centre of the hyperbola. Then N is a fixed point and ON a fixed line;

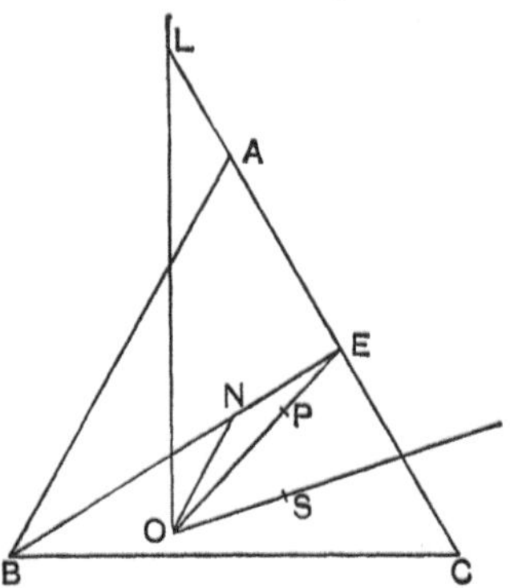

also since O lies on the nine-point circle of the triangle, and N is the centre of the latter,

$$ON = NE = \tfrac{1}{2}R.$$

Let OL be the asymptote, and S the focus of the hyperbola; let $NOS = \phi$. Then if OE meet the hyperbola in P, $OE = EL$; and from the equation of a rectangular hyperbola referred to its asymptotes

$$OS^2 = 4OP^2 \sin LOP \cos LOP = 2OP^2 \sin OEC.$$

Also from the equation referred to a pair of conjugate diameters

$$AE^2 = OE^2 - OP^2 ;$$

but

$$AE = \tfrac{1}{2}R\sqrt{3}, \quad OE = 2ON \cos NEO = R \sin OEC,$$

whence

$$OS^2 = \tfrac{1}{2}R^2 \left(4 \sin^3 OEC - 3 \sin OEC\right)$$
$$= -\tfrac{1}{2}R^2 \sin 3OEC.$$

But

$$OEC = 2LOE = 2\left(\tfrac{3}{4}\pi - \phi - OEC\right),$$

whence

$$OS^2 = \tfrac{1}{2}R^2 \sin 2\phi,$$

and therefore the locus of S is a lemniscate.

256. *To find the equation of the evolute of a lemniscate.*

In the figure to § 253, let O be the centre of curvature at P, Q the corresponding point on the rectangular hyperbola which is the first negative pedal of the lemniscate. Let the normal at P meet CQ in K. Then

$$\theta = QCA = ACP = \tfrac{1}{2}CPK,$$

whence

$$KC = KP = KQ.$$

Let $CQ = r'$, then by § 253,

$$\rho = \tfrac{1}{3}r', \text{ and } OK = \tfrac{1}{6}r',$$

whence if (x, y) be coordinates of O,

$$x = \tfrac{1}{2}r' \cos \theta + \tfrac{1}{6}r' \cos 3\theta = \tfrac{2}{3}r' \cos^3 \theta,$$

$$y = \tfrac{1}{2}r' \sin \theta - \tfrac{1}{6}r' \sin 3\theta = \tfrac{2}{3}r' \sin^3 \theta,$$

whence
$$x^{\frac{2}{3}} + y^{\frac{2}{3}} = (\tfrac{2}{3}r')^{\frac{2}{3}};$$

also
$$x^{\frac{4}{3}} - y^{\frac{4}{3}} = (\tfrac{2}{3}r')^{\frac{4}{3}} (\cos^4 \theta - \sin^4 \theta)$$
$$= (\tfrac{2}{3}r')^{\frac{4}{3}} \cos 2\theta,$$

whence
$$(x^{\frac{2}{3}} + y^{\frac{2}{3}})(x^{\frac{4}{3}} - y^{\frac{4}{3}}) = \tfrac{4}{9}r'^2 \cos 2\theta$$
$$= \tfrac{4}{9}a^2,$$

which is the equation of the evolute.

257. The lemniscate of Bernoulli, being the pedal of a rectangular hyperbola with respect to its centre, belongs to the class of curves included in the equation

$$(x^2 + y^2)^2 = a^2x^2 + b^2y^2,$$

which is the pedal and also the inverse of a central conic with respect to its centre. These curves are trinodal quartics having a pair of ordinary nodes at the circular points, and a biflecnode at the origin which will be complex or real according as the conic is an ellipse or hyperbola. They are also included in the equation $x + \iota y = c \sec (\xi + \iota\eta)$, and are one of the few classes of curves whose potentials can be completely investigated. The two points which are the inverses of the foci of the conic, and also the biflecnode at the origin, are double foci. It also follows that if x and y are tangential coordinates, the preceding equation represents the first negative pedal of a central conic with respect to its centre; hence Plücker's numbers for the pedal are $n = 6$, $\delta = 4$, $\kappa = 6$, $m = 4$, $\tau = 3$, $\iota = 0$.

The Lemniscate of Gerono.

258. This curve has been sometimes confounded with Bernoulli's lemniscate, owing to its form being that of a figure of eight. It may be constructed as follows. Let P be any point on a circle whose centre is C and radius a; draw PM perpendicular

to any diameter CA, and PN perpendicular to the tangent at A. Join CN and let it intersect PM at Q. Then the locus of Q is the curve in question, and its equation is

$$x^4 = a^2(x^2 - y^2) \quad\quad\dots\dots\dots\dots\dots\dots(1).$$

The curve has a biflecnode at the origin, and a tacnode at infinity, and therefore belongs to species VII. To prove the latter statement, transform to trilinear coordinates so that the axes of x and y are the sides BC, BA, whilst the line at infinity is the third side of the triangle of reference; then (1) becomes

$$\gamma^4 = a^2(\gamma^2 - \alpha^2)\beta^2 \dots\dots\dots\dots\dots\dots(2).$$

Now if in (16) of § 165 we interchange β and γ, the resulting equation represents a quartic having a tacnode at A and the line $\beta = 0$ or AC as the tacnodal tangent; and if in the resulting equation we put $\lambda = \mu = 0$, $v_1 = 0$, $v_2 = - v_0\gamma^2$, it reduces to (2).

The Oval of Descartes.

259. The oval of Descartes is the locus of a point P which moves so that its distances from two fixed points F, F_1 are connected by the relation

$$FP + mF_1P = a,$$

where m and a are constants.

The two points F, F_1, as well as a third point F_2 (see § 262) will be provisionally called the foci; and we shall prove in § 273 that these three points satisfy Plücker's definition of foci.

Let $FP = r$, $F_1P = r_1$, $FF_1 = c$, $PFF_1 = \theta$, then the polar equation of the curve is

$$r^2(1 - m^2) - 2r(a - m^2c\cos\theta) + a^2 - m^2c^2 = 0\dots\dots(1).$$

If this equation is written in the form

$$\{r^2(1 - m^2) + 2m^2cx + a^2 - m^2c^2\}^2 = 4a^2r^2$$

it is identical with what (12) of § 200 becomes when $a = b$; and is therefore a cartesian.

If the curve be defined by the equation

$$r + mr_1 = a \quad\dots\dots\dots\dots\dots\dots\dots(2),$$

it follows that r, r_1 and c are essentially positive quantities, whilst m and a may have any positive or negative values subject to the condition that (2) should represent a real curve. If, on the other hand, the curve be defined by the polar equation (1), c may be negative, in which case we shall obtain a more general species of curves which possess two cusps at the circular points, and are therefore *cartesians*, but which cannot be generated by Descartes' method.

260. In order that the vectorial coordinates r, r_1 should represent a *real* point, it is necessary that the circles whose centres are F, F_1 and radii r, r_1 should cut one another; this requires that

$$r + c > r_1 > r - c.$$

We shall now show how to determine the limiting values of m and a in order that the curve may be real*.

Let OX, OY be two rectangular axes; $OA = OB = c$; draw AP, Bp perpendicular to AB. Let r, r_1 be vectorial coordinates of a point referred to F, F_1 as foci, where $FF_1 = c$; and let $x = r$, $y = r_1$ be the coordinates of a point Q in the plane XOY referred to OX, OY as axes.

The condition that Q should lie within the rectangle $PABp$ is that $pM > QM > PM$; but

$$pM = A'M = r + c,$$
$$PM = AM = r - c,$$

accordingly the condition becomes

$$r + c > r_1 > r - c.$$

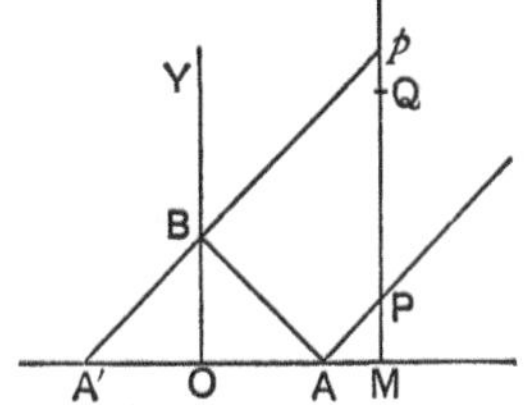

Hence all points lying within the rectangle $PABp$ correspond to real points in vectorial coordinates; and all points lying outside it to imaginary points. Accordingly the condition that the equation $r + mr_1 = a$ should represent a real curve is the same as the condition that the straight line $x + my = a$ should cut the rectangle.

Let this straight line cut the axes in D and E; then there will be four cases to consider according as D lies (i) on OA produced, (ii) between O and A, (iii) between O and A', (iv) on OA' pro-

* Crofton, *Proc. Lond. Math. Soc.* vol. I. p. 5.
S. Roberts, *Ibid.* vol. III. p. 106.
Cayley, *Ibid.* p. 181.

duced. Recollecting that $OD = a$, $OE = a/m$, we obtain the following results :—

Case I. $a > c$; $\infty > m > -1$.

Case II. $c > a > 0$; $a/c > m > -\infty$.

Case III. $0 > a > -c$; $a/c > m > -\infty$.

Case IV. $-c > a > -\infty$; $-1 > m > -\infty$.

If a or m lie outside these limits, the curve is imaginary.

Cases I. and IV.

261. We shall now examine the geometrical meaning of these conditions, and we shall find it convenient to begin with the two ovals

$$r + mr_1 = a \dots\dots\dots\dots\dots\dots(3),$$

$$r - mr_1 = a \dots\dots\dots\dots\dots\dots(4),$$

in which $a > c$ and $1 > m > 0$.

With F as a centre describe a circle whose radius $FR = a$. On FR take two points P and Q such that

$$\frac{PR}{F_1P} = \frac{QR}{F_1Q} = m.$$

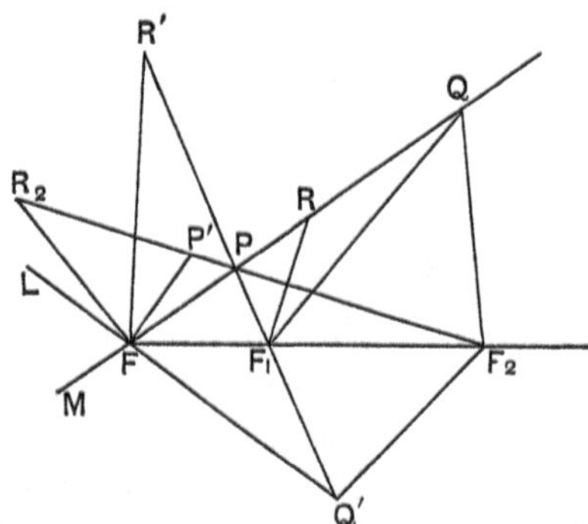

Then P will be a point on (3) and Q a point on (4); also F_1R bisects the angle PF_1Q. Let

$$FF_1R = \phi, \quad PF_1R = \alpha, \quad FRF_1 = R;$$

then $FP = \dfrac{c \sin(\phi - \alpha)}{\sin(R + \alpha)}, \quad FQ = \dfrac{c \sin(\phi + \alpha)}{\sin(R - \alpha)},$

whence $FP . FQ = \dfrac{c^2 (\sin^2 \phi - \sin^2 \alpha)}{\sin^2 R - \sin^2 \alpha}.$

But
$$\frac{\sin \alpha}{\sin R} = \frac{PR}{F_1 P} = m,$$

and
$$\frac{\sin \phi}{\sin R} = \frac{a}{c},$$

whence
$$FP \cdot FQ = \frac{a^2 - m^2 c^2}{1 - m^2}.$$

Accordingly the rectangle $FP \cdot FQ$ is constant; if therefore the circle circumscribing PQF_1 cut FF_1 in F_2, and $FF_2 = c'$, it follows that

$$FP \cdot FQ = FF_1 \cdot FF_2 \quad \dotfill (5),$$

$$c' = FF_2 = \frac{a^2 - m^2 c^2}{c(1 - m^2)} \quad \dotfill (6),$$

and consequently F_2 is a fixed point.

262. *To prove that F_2 is the third focus of the curve.*

Let $F_2 P = r_2$; then since the triangles $F_2 FP$ and QFF_1 are similar

$$\frac{r_2}{c'} = \frac{F_1 Q}{FQ} = \frac{QR}{mFQ} = \frac{FQ - a}{mFQ}.$$

But $r \cdot FQ = cc'$, whence

$$r + r_2 mc/a = cc'/a \dotfill (7).$$

Eliminating r between (3) and (7) we get

$$r_2 c/a - r_1 = (cc' - a^2)/ma \dotfill (8).$$

From (7) and (8) it follows that a relation similar to (2) exists between FP and $F_2 P$, and also between $F_1 P$ and $F_2 P$; whence F_2 is a third focus of the inner oval P. In the same way it can be shown that F_2 is a third focus of the outer oval Q.

263. *If PF_1 be produced to meet the outer oval in Q', then*
$$F_1 P \cdot F_1 Q' = F_1 F \cdot F_1 F_2.$$

Produce $F_1 P$ to R' so that $F_1 R' = a/m$; then since

$$F_1 P + PR' = a/m,$$

and
$$FP + mF_1 P = a,$$

we obtain
$$mPR' = FP.$$

Similarly
$$mQ'R' = FQ',$$

whence
$$\frac{FQ'}{Q'R'} = \frac{FP}{PR'} = m,$$

and therefore FR' bisects the external angle PFL. Let

$$F_1FR = \phi, \quad PFR' = R'FL = \alpha;$$

then proceeding in exactly the same way as in § 261, it can be shown that

$$F_1P \cdot F_1Q' = \frac{a^2 - c^2}{1 - m^2} = F_1F \cdot F_1F_2 \dots\dots\dots\dots(9).$$

Hence the circle circumscribing PFQ' passes through F_2.

264. *If F_2P be produced to meet the inner oval in P', then*

$$F_2P \cdot F_2P' = F_2F_1 \cdot F_2F.$$

Produce F_2P to R_2 so that $F_2R_2 = c'/m$; then

$$F_2P + PR_2 = c'/m,$$

also by (7) $\qquad F_2P + FP \cdot a/mc = c'/m,$

whence $\qquad PR_2 = FP \cdot a/mc.$

Similarly $\qquad P'R_2 = FP' \cdot a/mc,$

whence $\qquad \dfrac{FP}{PR_2} = \dfrac{FP'}{P'R_2} = \dfrac{mc}{a},$

and therefore FR_2 bisects the external angle $P'FM$. Putting

$$F_2FR_2 = \phi, \quad P'FR_2 = R_2FM = \alpha,$$

and proceeding as before, we obtain

$$F_2P \cdot F_2P' = \frac{a^2 - c^2}{a^2 - m^2c^2} = F_2F_1 \cdot F_2F \dots\dots\dots\dots(10).$$

265. We must now examine the positions of the foci with respect to the curve.

Let (3) cut the line FF_1 in A and B, where A lies on the left of F; then

$$FA = \frac{a - mc}{1 + m}, \quad F_1B = \frac{a - c}{1 + m}, \quad AB = \frac{2a}{1 + m}.$$

Since $a > c$ and $m < 1$, it follows that FA and F_1B are both positive, and therefore the foci F_1F_2 lie inside the oval (3).

The only values of r and r_1 which simultaneously satisfy (3) and (4) are $r = a$, $r_1 = 0$; but when $r_1 = 0$, $r = c$, and since $a > c$ this is impossible, and therefore the ovals do not intersect; a result which might be foreseen from the fact that a quartic cannot have four double points. Whence the oval (3) lies inside the oval (4), and the foci F and F_1 lie within both ovals.

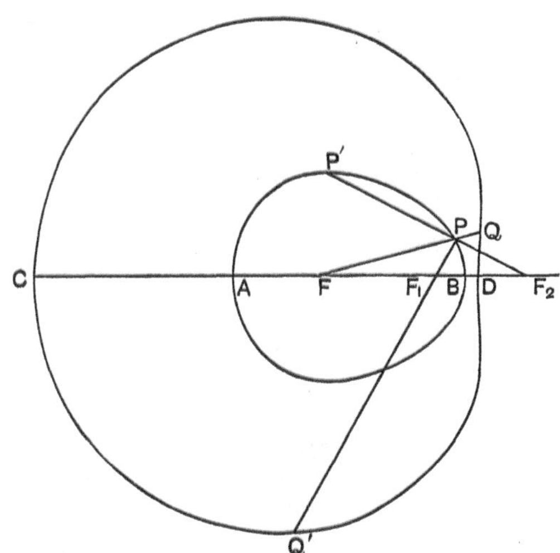

Let the oval (4) cut FF_1 in C and D, where C lies on the left of F: then

$$FD = \frac{a - mc}{1 - m}, \quad FF_2 - FD = \frac{(a - mc)(a - c)}{c(1 - m^2)},$$

whence $FF_2 > FD$ and therefore F_2 lies outside both ovals.

The three equations (5), (9) and (10) are fundamental ones in the theory of these curves. The first one shows that each oval is the inverse of the other with respect to the internal focus F; the second shows that either oval turned through two right angles is the inverse of the other with regard to the central focus F_1; whilst the third shows that each oval is its own inverse with respect to the external focus F_2. Also the two tangents drawn from F_2 to the inner and outer ovals respectively are equal.

A pair of ovals which possess these focal properties are called *conjugate ovals*; and their forms when $a > c$, $m < 1$ are shown in the figure.

266. When $m = 1$, $F_1P = PR$, see figure to § 261, whence the angle

$$PRF = PF_1R = RF_1Q;$$

accordingly F_1Q is parallel to FP, also from (6) $FF_2 = \infty$. In this case every point of the outer oval and also the external focus move off to infinity, whilst the inner oval becomes an ellipse whose foci are F and F_1 and whose major axis is a.

When $m > 1$ and $a > mc$, the oval (4) becomes imaginary, but the line F_1Q now cuts PF produced so that F_1R bisects the external angle of the triangle F_1PQ; hence

$$\frac{QR}{F_1Q} = \frac{PR}{F_1P} = m,$$

also since

$$FQ + FR = QR = mF_1Q,$$

the locus of Q is the oval

$$r + a = mr_1 \quad \dots\dots\dots\dots\dots\dots(11).$$

Writing (3) and (11) in the forms

$$\left.\begin{array}{l} r_1 + r/m = a/m \\ r_1 - r/m = a/m \end{array}\right\} \quad \dots\dots\dots\dots\dots\dots(12),$$

it follows that if $a > mc$, these ovals are of the same species as the pair of conjugate ovals we have previously been discussing, but the foci F and F_1 are interchanged. Also writing (11) in the form $r - mr_1 = -a$, it follows that (11) belongs to Case IV., in which a and m are negative quantities which are numerically greater than c and 1 respectively.

From (6) it follows that FF_2 is negative, so that F_2 lies on the left of F, and its value is $(a^2 - m^2c^2)/c\,(m^2 - 1)$.

When $m = a/c$, $FF_2 = 0$, and F_2 coincides with F; also both ovals pass through F since (12) are satisfied by $r_1 = c$, $r = 0$. The polar equations of the ovals referred to the focus F are

$$r\,(a^2 - c^2) = 2ac\,(a \cos \theta \mp c),$$

the upper and lower signs being used for the internal and external ovals respectively. But when polar coordinates are employed negative as well as positive values of r are admissible, whence both ovals are included in the equation

$$r\,(a^2 - c^2) = 2ac\,(a \cos \theta - c) \quad \dots\dots\dots\dots(13),$$

which is a *hyperbolic limaçon* whose node is at F, which is also a double focus. The focus F_1 lies inside the internal loop.

Lastly let $m > a/c$; then from (6) c' is positive, and therefore F_2 lies on the right of F; but $FF_1 > FF_2$, so that F_2 now becomes the middle and F the external focus. To find the conjugate oval, produce F_1P to R so that $F_1R = a/m$, and on F_1R take a point Q such that angle $QFR = RFP$; then since by (3) the locus of P may be written $F_1P + FP/m = a/m$, it follows that $RP = FP/m$, whence

$$\frac{RP}{FP} = \frac{RQ}{FQ} = \frac{1}{m},$$

and

$$F_1Q - FQ/m = F_1R = a/m,$$

and therefore the locus of Q is the oval

$$r - mr_1 = - a \dots\dots\dots\dots\dots\dots\dots(14).$$

It can also be shown as in § 261 that

$$F_1P \cdot F_1Q = \frac{a^2 - c^2}{m^2 - 1} = F_1F_2 \cdot F_1F,$$

and consequently Q is a point on the conjugate oval, which by (14) belongs to Case IV.

Cases II. and III.

267.   In Case II, $c > a > 0$, $a/c > m > - \infty$; but we shall find it convenient to begin by discussing the two ovals

$$r - mr_1 = a\dots\dots\dots\dots\dots\dots\dots\dots(15),$$

$$r + mr_1 = a\dots\dots\dots\dots\dots\dots\dots\dots(16),$$

in which m is a positive quantity lying between a/c and 0.

In the figure $FR = a$, whilst P and Q are two points such that

$$\frac{RP}{F_1P} = \frac{RQ}{F_1Q} = m,$$

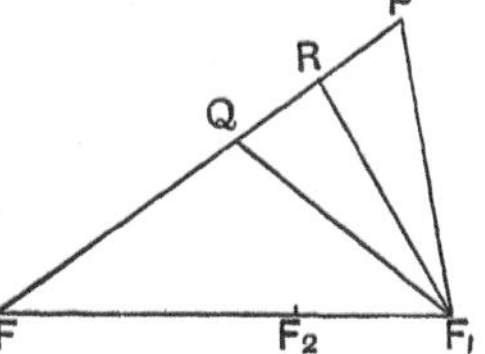

whence (15) and (16) are the equations of the loci of P and Q respectively.   We can also prove as before that

$$FQ \cdot FP = FF_2 \cdot FF_1 = cc' = \frac{a^2 - m^2c^2}{1 - m^2};$$

accordingly as long as m lies between 0 and a/c, the value of F_2 is positive and less than c, whence F_2 is the central focus and F_1 the external one.

When $m = 0$, both ovals coalesce into a circle of radius a whose inverse points are F_2, F_1.   When $mc = a$, $FF_2 = 0$ and

12—2

F_2 as well as Q coincide with F; and the locus of P is the *elliptic limaçon*

$$r\,(c^2 - a^2) = 2ac\,(c - a\cos\theta),$$

the two foci of which are F and F_1. The point F is a conjugate point and also a double focus.

When $1 > m > a/c$, the value of c' becomes negative and F_2 lies to the left of F, so that F now becomes the middle focus and the oval (16) becomes imaginary. To find the conjugate oval, we observe that Q now lies on PF produced, whence it can be shown as before that the locus of Q is the conjugate oval

$$r - mr_1 = -a \quad\dots\dots\dots\dots\dots\dots\dots(17),$$

which is one of the curves belonging to Case III.

When $m = 1$, F_2 moves off to $-\infty$, whilst (15) becomes the right-hand branch and (17) becomes the left-hand one of a hyperbola, whose foci are F and F_1, and whose major axis is a.

When $m > 1$, c' and also $c' - c$ are both positive, and therefore F_2 lies on the right of F_1 and is therefore the external focus.

The locus of P is given by (15), which may be written

$$r_1 + a/m = r/m\dots\dots\dots\dots\dots\dots\dots(18).$$

To find the conjugate oval, produce PF_1 to R so that $F_1R = a/m$; draw FQ so that the angle $QFR = RFP$. Then by (18), $FP/m = F_1P + F_1R = PR$; whence

$$m = \frac{FP}{PR} = \frac{FQ}{QR};$$

accordingly the locus of Q is

$$FQ - mF_1Q = -a,$$

which belongs to Case III.

Also if $F_1FR = \phi$, $PFR = RFQ = \alpha$, it can be shown as before that

$$F_1P \cdot F_2Q = \frac{c^2\,(c^2 - a^2)}{m^2 - 1},$$

and therefore Q is a point on the conjugate oval.

The oval of Descartes belongs to species VI, for which Plücker's numbers are $n = 4$, $\delta = 0$, $\kappa = 2$, $m = 6$, $\tau = 1$, $\iota = 8$; whilst the two limaçons belong to species IX, for which $n = 4$, $\delta = 1$, $\kappa = 2$, $m = 4$, $\tau = 1$, $\iota = 2$.

Having classified the oval of Descartes, we shall add a few properties of the curve.

268. *If a radius vector be drawn from the focus F cutting two conjugate ovals in P and Q, the tangents at P and Q intersect at the middle point of the arc PQ of the circle passing through PQF_2F_1.*

If the equation of the inner and outer ovals be

$$r \pm mr_1 = a,$$

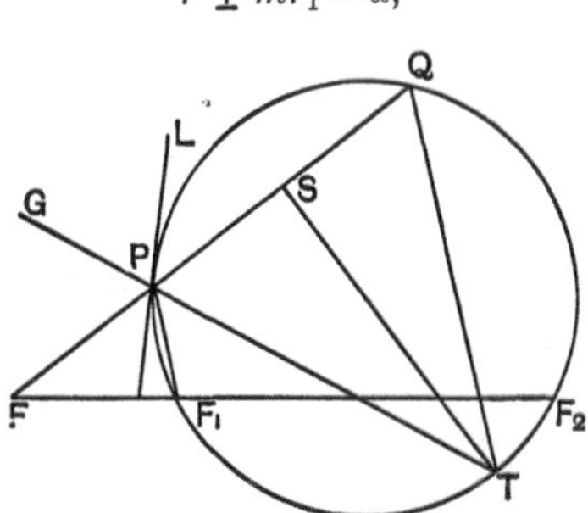

we obtain $\qquad dr/ds \pm mdr_1/ds = 0,$

whence $\qquad \cos \phi = m \cos \phi_1.$

We have shown that $FP \cdot FQ = FF_1 \cdot FF_2$, whence a circle can be described round PQF_2F_1. Let S be the middle point of PQ; draw ST at right angles to PQ cutting the circle in T; join TP, TQ. Let $FPG = \phi$, $F_1PT = \phi_1$; then

$$\phi = TPQ = TQP,$$
$$\phi_1 = TPF_1 = TQF_1,$$

whence $\qquad F_1PQ = \phi + \phi_1, \quad F_1QP = \phi - \phi_1.$

Now $\qquad FP + mF_1P = a = FQ - mF_1Q,$

whence

$$m\,(F_1P + F_1Q) = PQ = F_1Q \cos(\phi - \phi_1) + F_1P \cos(\phi + \phi_1).$$

But

$$\frac{F_1P}{\sin(\phi - \phi_1)} = \frac{F_1Q}{\sin(\phi + \phi_1)};$$

substituting and reducing we get

$$\cos \phi = m \cos \phi_1,$$

which shows that TP is the tangent at P. Similarly TQ is the tangent at Q.

269. *The tangent at P bisects the angle between the focal distance FP and the tangent at P to the circle through F_1PF_2.*

If PL be the tangent to the circle,

$$FPG = TPQ = TQP = GPL.$$

270. *The locus of S the middle point of PQ is a limaçon.*

The polar equation, when F is the origin, is given by (1), whence

$$FS = \tfrac{1}{2}\,(r_1 + r_2) = (a - m^2 c \cos \theta)/(1 - m^2).$$

All the preceding propositions hold good for either of the other two foci, provided P and Q are points satisfying the focal properties (9) and (10). The footnote* contains a list of some recent memoirs on this curve.

271. *If any chord cut a cartesian in four points, the sum of its distances from any focus is constant.*

The equation of a cartesian referred to any focus is

$$r^2 + 2r\,(a + b \cos \theta) + \delta^2 = 0\,;$$

let the equation of any straight line be $r\,(A \cos \theta + B \sin \theta) = 1$; then if we eliminate θ between these two equations, we shall obtain a quartic equation for r, in which the coefficient of r^3 is equal to $4a$.

272. A cartesian has eight points of inflexion, and since the curvature at such points vanishes and changes sign, the radius of curvature becomes infinite at a point of inflexion. Hence the denominator of the expression for the radius of curvature, when equated to zero, furnishes the equation of the curve which passes through the points of inflexion; and in the case of a cartesian the curve is a circular cubic, whose equation may be found from that of the curve by equating the value of d^2y/dx^2 to zero.

The last two propositions are true for all cartesians.

* Genocchi, *Nouv. Ann.* 1855; *Mathesis*, 1884.
  Zeuthen, *Ibid.* 1864, p. 304.
  Sylvester, *Phil. Mag.* vol. XXXI. 1866.
  D'Ocagne, *Comp. Rend.* 1883, p. 1424.
  Liguine, *Bull. de Darboux*, 1882; *Interm. des Math.* 1896, p. 238.

Foci.

273. *To prove that the three points F, F_1, F_2 satisfy Plücker's definition of a focus.*

The equation of an oval of Descartes referred to the point F is

$$\{r^2(1-m^2)+2m^2cx+a^2-m^2c^2\}^2 = 4a^2r^2\ldots\ldots(19),$$

and the points where the line $x+\iota y=0$ intersects (19) are determined by the equation

$$(2m^2cx+a^2-m^2c^2)^2 = 0,$$

which shows that this line is a tangent to the curve. In the same way it can be shown that the line $x-\iota y=0$ is also a tangent; whence the point F satisfies Plücker's definition of a focus.

Since the polar equation of the curve referred to F_1 and F_2 is of the same form as (19), it follows that these points are also foci.

Since cartesians are bicuspidal quartics of the sixth class, it follows that these curves have one triple and three single foci. The latter have already been determined; the former may be obtained by considering the bicuspidal quartic

$$\{\beta\gamma(1-m^2)+m^2c(\beta+\gamma)\alpha+(a^2-m^2c^2)\alpha^2\} = 4a^2\alpha^2\beta\gamma,$$

which reduces to a cartesian when B and C are the circular points, and α the line at infinity. The two cuspidal tangents are

$$\beta(1-m^2)+m^2c\alpha = 0,$$
$$\gamma(1-m^2)+m^2c\alpha = 0,$$

which, when retransformed into Cartesian coordinates, become

$$(x \pm \iota y)(1-m^2)+m^2c = 0,$$

which intersect at the real point

$$x = -m^2c/(1-m^2), \quad y = 0 \ \ldots\ldots\ldots(20),$$

which is the triple focus.

If the origin be transferred to the triple focus, it will be found that the curve assumes the form $S^2+L=0$; whence the triple focus is the centre of the focal circle.

274. We have shown in § 267 that when $m = a/c$ the curve becomes a limaçon, and that two of the single foci coincide at the node, which becomes a double focus which agrees with § 212, whilst the single focus lies without or within the curve according as the limaçon is elliptic or hyperbolic. The distance of the triple focus from the node is $a^2c/(a^2 - c^2)$.

275. To find the corresponding results for a cardioid, put $a = c = (1 - m^2) A$ in (1) of § 259, divide out by $1 - m^2$, and then put $m = 1$, and we obtain

$$r = 2A (1 - \cos \theta).$$

Hence $c = 0$, and by (6) of § 261 $c' = 0$, accordingly the cusp is a triple focus, which agrees with § 212. The other triple focus lies within the curve and on the left-hand side of the cuspidal focus, from which its distance is equal to $-A$, or one-fourth of the distance of the cusp from the vertex.

276. We shall lastly consider the case of a cartesian with three collinear foci, two of which are imaginary.

Writing

$$f = \frac{m^2c}{1 - m^2}, \quad \delta^2 = \frac{a^2 - m^2c^2}{1 - m^2}, \quad A = \frac{a}{1 - m^2} \quad \ldots\ldots(21),$$

the equation of the curve may be written

$$(r^2 + 2fr + \delta^2)^2 = 4A^2r^2 \quad \ldots\ldots\ldots\ldots\ldots(22).$$

To determine the single foci, we must find the condition that the line $x - \alpha + \iota (y - \beta) = 0$ should touch the curve; whence putting $\alpha + \iota\beta = p$, and eliminating y from (22), we obtain

$$4x^2 (p + f)^2 - 4x \{(p + f) (p^2 - \delta^2) - 2A^2p\} + (p^2 - \delta^2)^2 + 4A^2p^2 = 0,$$

and the condition that the roots of this quadratic should be equal is that

$$p \{p^2f + p (\delta^2 + f^2 - A^2) + f\delta^2\} = 0.$$

The factor $p = 0$ gives $\alpha = 0$, $\beta = 0$ which determines the origin, which by hypothesis is the real single focus; whilst the other factor determines the remaining single foci. Now if the roots of the equation

$$p^2f + p (\delta^2 + f^2 - A^2) + f\delta^2 = 0 \ldots\ldots\ldots\ldots(23)$$

be real, we must have $\beta = 0$, $\alpha = p$, in which case there will be a pair of real single foci on the axis of x. It will also be found that if m and a be eliminated from (21), the result is (23) with c substituted for p. If however the roots of (23) are complex and equal to $P \pm \iota Q$, the equations of the two tangents drawn from one of the circular points are

$$x + \iota y = P + \iota Q, \quad x + \iota y = P - \iota Q,$$

whilst the equations of the two tangents drawn from the other circular point are

$$x - \iota y = P + \iota Q, \quad x - \iota y = P - \iota Q.$$

These four straight lines intersect in the points

$$x = P + \iota Q, \quad y = 0,$$
$$x = P, \quad y = Q,$$
$$x = P, \quad y = -Q,$$
$$x = P - \iota Q, \quad y = 0.$$

Hence there are two imaginary foci which lie on the axis of x, and two real ones which are determined by the equations

$$x = P, \quad y = \pm Q.$$

The latter foci together with the origin are the only real single foci which the curve possesses.

277.   The coordinates of the points where (22) cuts the axis of x are determined by the equation

$$\{x^2 + 2(f + A)x + \delta^2\}\{x^2 + 2(f - A)x + \delta^2\} = 0,$$

and the condition that the values of x, obtained by equating *both* factors to zero, should be real is that

$$(\delta + f + A)(\delta - f - A) \text{ and } (\delta + f - A)(\delta - f + A) \quad \ldots(24)$$

should be both negative.   Now the condition that the roots of (23) should be real is that

$$(\delta + f + A)(\delta + f - A)(\delta - f + A)(\delta - f - A)$$

should be positive.   Hence it follows from (24) that when the three collinear foci are all real the curve cuts the axis in four real points, but when two of these foci are imaginary, the curve cuts the axis in two real and two imaginary points.

Two of the forms of the curve in the latter case are shown in the figure, and further information will be found in Cayley's *Memoir on Caustics**.

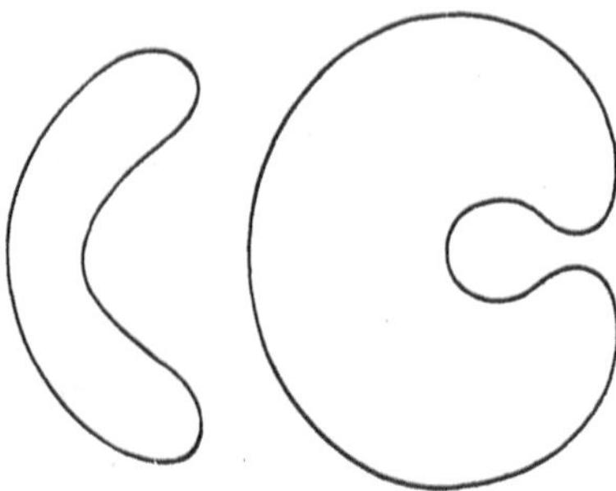

The Limaçon.

278. We have shown in § 266 that a limaçon is a particular case of an oval of Descartes in which two of the foci coincide. It is, however, more usual to define this curve as the inverse of a conic with respect to its focus. The polar equation of a conic is

$$l/r = 1 - e \cos \theta,$$

whence the polar equation of a limaçon is

$$r = a - b \cos \theta \quad \dots\dots\dots\dots\dots\dots\dots(1),$$

where $b/a = e$. The curve is therefore the inverse of an ellipse or a hyperbola according as $a >$ or $< b$; in the former case it is called an *elliptic limaçon* and in the latter a *hyperbolic limaçon*. This curve appears to have been first studied by Pascal, who so named it from a fancied resemblance to the form of a snail. When $a = b$, the curve is the inverse of a parabola with respect to its focus and is called a cardioid.

The Elliptic Limaçon.

279. The form of the elliptic limaçon is shown in the figure. The origin F is the point where the two internal foci of an oval of Descartes unite, and is also a conjugate point; whilst the external focus F_1 is the inverse of the other focus of the ellipse. Since the limaçon has a node at the origin and a pair of imaginary cusps at

* *Phil. Trans.* 1857, p. 273; *Collected Papers*, vol. II. p. 336.

the circular points, it belongs to the ninth species of quartics for which Plücker's numbers are $n = 4$, $\delta = 1$, $\kappa = 2$, $m = 4$, $\tau = 1$, $\iota = 2$, $D = 0$, and therefore has a triple focus, which is the point of intersection of the cuspidal tangents at the circular points.

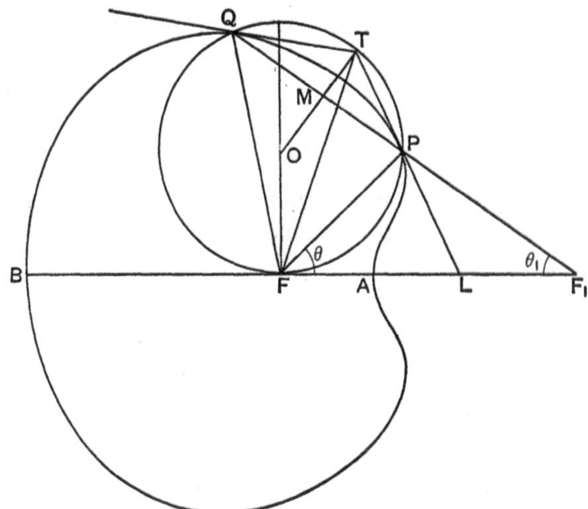

280.   *To find the position of the triple focus.*

Transform (1) into trilinear coordinates by writing

$$\beta = x + \iota y, \quad \gamma = x - \iota y, \quad I = 1,$$

and it becomes $\{\beta\gamma + \tfrac{1}{2}b\,(\beta + \gamma)\,I\}^2 = aI^2\beta\gamma.$

The cuspidal tangents at the circular points (β, I) and (γ, I) are $2\gamma + bI = 0$ and $2\beta + bI = 0$, or in Cartesian coordinates $x - \iota y + \tfrac{1}{2}b = 0$, $x + \iota y + \tfrac{1}{2}b = 0$, which intersect at the point $x = -\tfrac{1}{2}b$, $y = 0$, which is therefore the triple focus.

From (1) we obtain

$$FA = a - b, \quad FB = a + b, \quad AB = 2a,$$

from which it follows that if a point A' be taken on the opposite side of F such that $FA' = FA$, and S be the triple focus, $FS = \tfrac{1}{4}A'B$. Also the distance FF_1 can easily be shown to be equal to $\tfrac{1}{2}(a^2 - b^2)/b$, from which it follows that $AF_1 = (a - b)^2/2b$, and is therefore positive; whence F_1 lies outside the curve. The vectorial equation of the curve is

$$ar - br_1 = \tfrac{1}{2}(a^2 - b^2)\dots\dots\dots\dots\dots\dots(2),$$

which shows that F and F_1 possess the properties of foci.

281. Putting $F_1P = r_1$, $PF_1F = \theta_1$, $\frac{1}{2}(a^2 - b^2)/b = f = FF_1$, the polar equation of the curve referred to F_1 is

$$r_1{}^2 - r_1\,(b^2 + a^2 \cos \theta_1)/b + f^2 = 0 \dots\dots\dots\dots(3).$$

Whence if F_1PQ be any chord, the locus of the middle point of PQ is the hyperbolic limaçon

$$r = \tfrac{1}{2}(b^2 + a^2 \cos \theta_1)/b \dots\dots\dots\dots\dots(4),$$

also
$$F_1P \,.\, F_1Q = F_1F^2 \dots\dots\dots\dots\dots(5).$$

Equation (5) shows (i) that the curve is its own inverse with respect to the external focus F_1, which is therefore a centre of inversion; (ii) that the triangles F_1QF and FF_1P are similar; (iii) that the circle circumscribing FPQ touches FF_1 at F. Also from the properties of inverse curves, the angles TPQ and TQP, made by the tangents at P and Q with F_1P, are equal.

282. Let the tangent at P meet FF_1 in L; let $FPL = \phi$, $PLF_1 = \psi$; then

$$\tan \phi = (a - b \cos \theta)/b \sin \theta,$$

$$\tan \psi = \frac{a \cos \theta - b \cos 2\theta}{b \sin 2\theta - a \sin \theta}.$$

The form of the curve shows that at the point of contact of the double tangent, $\phi = \tfrac{1}{2}\pi - \theta$, whence

$$\cos \theta = \tfrac{1}{2}a/b, \quad r = \tfrac{1}{2}a \dots\dots\dots\dots(6).$$

Accordingly the points of contact of the double tangent will be real if $a < 2b$.

Making ψ a minimum we obtain

$$\cos \theta = (a^2 + 2b^2)/3ab \dots\dots\dots\dots(7),$$

which determines the two points of inflexion. In order that they may be real it is necessary that $2b > a > b$. When $a = 2b$, the vertex A is a point of undulation.

283. The Cartesian equation of a limaçon is

$$(x^2 + y^2 + bx)^2 = a^2\,(x^2 + y^2),$$

which shows that the origin is a conjugate point or a crunode according as the limaçon is elliptic or hyperbolic; also since the curve is of the ninth species, its reciprocal polar is another quartic of the same species. When $2b > a > b$, the reciprocal curve has two real cusps, one crunode, a real double tangent touching the curve at two imaginary points, and two imaginary points of inflexion. When $2b = a$, the vertex A becomes a point of undulation, and the reciprocal singularity is a triple point composed of a crunode and two real cusps. When $a > 2b$ the points of inflexion on the limaçon are imaginary, and the double tangent touches the curve at two imaginary points; hence the reciprocal curve has two imaginary cusps and a conjugate point.

When $a < b$, the limaçon is hyperbolic, and has two imaginary points of inflexion and a double tangent touching it at two real points. Hence the reciprocal curve has a crunode, two imaginary cusps, a double tangent touching the curve at two real points, and two imaginary points of inflexion.

The reader will find it an instructive exercise to trace the form of the reciprocal curve when the origin of reciprocation moves along the axis of x from plus to minus infinity. When the limaçon is elliptic and has two real points of inflexion, the form of the reciprocal curve, when the origin lies between the vertex B and the point of intersection of the two stationary tangents, resembles that of figure 5 of § 159.

284. *The limaçon is the pedal of a circle with respect to any point in its plane.*

Let O be the point, C the centre of the circle; and draw OZ perpendicular to the tangent at any point P on the circle. Let $CP = a$, $CO = b$, $PCO = \theta$. Then

$$OY = a - b \cos \theta,$$

whence the locus of Y is an elliptic or hyperbolic limaçon according as O lies within or without the circle. When O lies on the circle, $a = b$, and the pedal is a cardioid.

285. *If T be the middle point of the arc PQ of the circle circumscribing FPQ (see fig. to § 279), then TP, TQ are the tangents at P and Q.*

Let $FPL = \phi$, $F_1PL = \phi_1$; then differentiating (2) with respect to s we obtain

$$a \cos \phi = b \cos \phi_1 \dots\dots\dots\dots\dots\dots\dots(8),$$

which gives the relation between the angles which the tangent makes with the two focal distances; whence the theorem can be proved in the same manner as the corresponding property of the oval of Descartes given in § 268.

286. *Tangents at the extremities of a chord through the external focus subtend equal angles at the internal focus; also the locus of their point of intersection is a cissoid.*

The first part follows at once, since

$$TFP = TQP = TPQ = TFQ.$$

To prove the second part, let $TFF_1 = \chi$; then, since OT is perpendicular to PQ,

$$\chi = \tfrac{1}{2}\pi - OFT = \tfrac{1}{2}FOT = \tfrac{1}{2}\pi - \tfrac{1}{2}\theta_1;$$

whence, if (x, y) be the coordinates of T,

$$\frac{x^2 + y^2}{y^2} = \operatorname{cosec}^2 \chi = \sec^2 \tfrac{1}{2}\theta_1 = \frac{2}{1 + \cos \theta_1} \quad \ldots\ldots\ldots (9).$$

Let M be the middle point of PQ, then

$$OM \sin \theta_1 + F_1 M \cos \theta_1 = f = (a^2 - b^2)/2b;$$

also, by (4), $\qquad\qquad F_1 M = (a^2 \cos \theta_1 + b^2)/2b,$

whence $\qquad\qquad OM = \dfrac{a^2 \sin^2 \theta_1 - b^2 (1 + \cos \theta_1)}{2b \sin \theta_1}.$

Also $\qquad\qquad OF = f \tan \theta_1 - OM \sec \theta_1$

$$= \frac{b (1 + \cos \theta_1)}{2 \sin \theta_1},$$

and $\qquad\qquad x = OF \sin \theta_1 = \tfrac{1}{2} b (1 + \cos \theta_1) \ldots\ldots\ldots\ldots (10),$

whence by (9) and (10), the locus of T is the cissoid

$$x (x^2 + y^2) = b y^2 \ldots\ldots\ldots\ldots\ldots\ldots (11).$$

287. *The locus of the point of intersection of two tangents at the extremities of a chord through the node is a nodal circular cubic.*

The equation of the limaçon in Cartesian coordinates is

$$(x^2 + y^2 + bx)^2 = a^2 (x^2 + y^2) \ldots\ldots\ldots\ldots (12).$$

Let (h, k) be any point; write down its polar cubic, transform to polar coordinates and then eliminate r by means of the polar equation of the curve, and we shall obtain

$$\{(a^2 - bh) \tan^2 \theta + 2bk \tan \theta + a^2 + b^2 + bh\}^2$$
$$= a^2 (k \tan \theta + 2b + h)^2 (1 + \tan^2 \theta),$$

which is the equation for determining the vectorial angles of the four tangents drawn from (h, k) to the curve. This may be written in the form

$$(a^2 - bh)\sin^2\theta + bk\sin^2\theta + (a^2 + b^2 + bh)\cos^2\theta$$
$$= a\{k\sin\theta + (2b + h)\cos\theta\}\ldots(13).$$

Let PFP' be the chord; then if θ be the vectorial angle of P, $\pi + \theta$ must be that of Q; whence, if (h, k) be the point of intersection of the tangents at P and P', (13) must be satisfied by θ and $\pi + \theta$. This requires that both sides of (13) should vanish, whence eliminating θ between the two equations formed by equating both sides of (13) to zero, we obtain

$$(a^2 - bh)(h + 2b)^2 + k^2(a^2 - 3b^2 - bh) = 0.$$

Transferring the origin to the point $-2b$, this becomes

$$bx(x^2 + y^2) = (a^2 + 2b^2)x^2 + (a^2 - b^2)y^2\ldots\ldots\ldots(14),$$

which is the inverse of a conic with respect to its vertex. When $a = b$, the limaçon becomes a cardioid and (14) reduces to the circle

$$x^2 + y^2 = 3bx,$$

the centre of which is the triple focus.

288. The form of (12) shows that the radius of the focal circle is equal to $\frac{1}{2}a$, and that the distance of its centre from the nodal focus is equal to $\frac{1}{2}b$. Since the radius δ of the fixed circle vanishes on account of the limaçon being the inverse of a conic, the theorem of § 206 becomes:—

If from the nodal focus F a line be drawn to meet the curve in P, and if FQ be drawn to meet the normal at P in Q, such that the angle $FPQ = PFQ$, the locus of Q is the focal circle.

Also if F_1 be the external focus, the theorem becomes:—

If from the external focus F_1 a chord F_1PQ be drawn, the normals at P and Q intersect on the focal circle corresponding to F_1.

289. *To find the p and r equation of the curve.*

We have, by the ordinary formulae,

$$p = r\sin\phi, \quad b\sin\theta = r\cot\phi,$$

whence
$$\frac{r^2}{p^2} = 1 + \frac{b^2\sin^2\theta}{r^2}.$$

Eliminating θ by (1), we get

$$r^4 = p^2 (b^2 - a^2 + 2ar) \quad\dots\dots\dots\dots\dots\dots(15)$$

$$= 2bp^2 r_1 \dots\dots\dots\dots\dots\dots\dots\dots\dots(16),$$

by (2), whence $\qquad \rho = r\dfrac{dr}{dp} = \dfrac{r^2(b^2 - a^2 + 2ar)}{p(2b^2 - 2a^2 + 3ar)}.$

It is shown in treatises on optics that the evolute of a limaçon is the caustic by reflexion of a circle. The evolute is a curve of the sixth degree*.

290.   *If a triangle be inscribed in a given circle, whose vertex A is fixed, and whose vertical angle A is constant, the locus of the centres of the inscribed and escribed circles is a limaçon.*

Let O be the centre of the inscribed circle, let $AO = r$, and let θ be the angle which AO makes with the tangent at A. Then, if R be the radius of the circumscribing circle, and r' that of the inscribed circle,

$$r' = 2R \sin \tfrac{1}{2}A \sin \tfrac{1}{2}B \sin \tfrac{1}{2}C;$$

also $\qquad\qquad r = r' \operatorname{cosec} \tfrac{1}{2}A$

$$= R\{\cos \tfrac{1}{2}(B - C) - \sin \tfrac{1}{2}A\}.$$

Also $\qquad\qquad \theta = C + \tfrac{1}{2}A = \tfrac{1}{2}\pi - \tfrac{1}{2}(B - C),$

whence $\qquad\qquad r = R\{\sin \theta - \sin \tfrac{1}{2}A\},$

and therefore the locus of O is a limaçon.

291.   It can also be proved: (i) that a limaçon is the locus of the vertex of a triangle whose sides slide on the circumferences of two given circles; (ii) that it is the epitrochoid generated by a point in the plane of a circle which rolls on another circle of equal radius.

The Hyperbolic Limaçon.

292.   When $a < b$, the equation $r = a - b\cos\theta$ represents a hyperbolic limaçon; and the form of the curve, which is shown in the figure, consists of an outer portion and a loop. We have

$$FA = b - a, \quad FB = b + a, \quad FF_1 = (b^2 - a^2)/2b.$$

---

* Heath, *Geometrical Optics*, p. 111.

Also since $F_1A = \frac{1}{2}(b-a)^2/b$, it follows that $FA > F_1A$, and therefore the focus F_1 lies inside the loop.

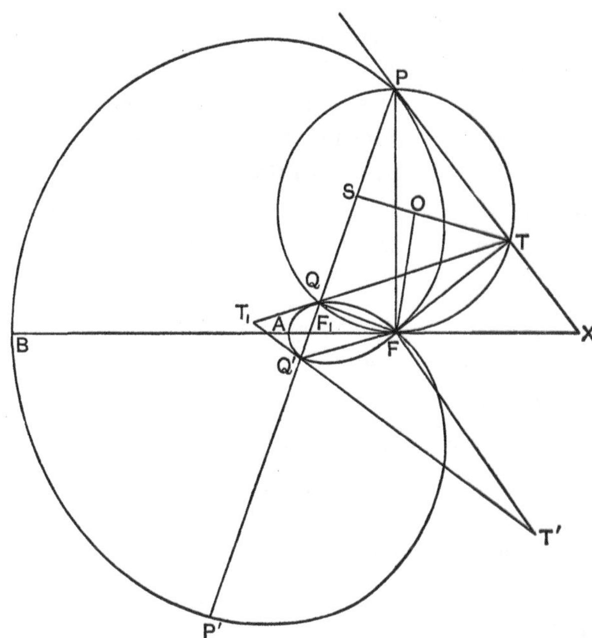

The two nodal tangents are inclined at angles $\pm\cos^{-1}a/b$ to the axis; also if P and Q are two points on the outer portion and the loop respectively, lying on the positive side of the axis, and if $PFA = \chi$, $QFA = \chi'$,

$$FP = a + b\cos\chi, \quad FQ = b\cos\chi' - a,$$

since the corresponding values of θ are $\theta = \pi - \chi$, $\theta = 2\pi - \chi'$.

The vectorial equation for a point P on the outer portion is

$$br_1 - ar = \tfrac{1}{2}(b^2 - a^2) \quad\ldots\ldots\ldots\ldots\ldots\ldots(1),$$

and for a point Q on the loop,

$$br_1 + ar = \tfrac{1}{2}(b^2 - a^2) \quad\ldots\ldots\ldots\ldots\ldots\ldots(2).$$

Also the polar equation referred to the internal focus F_1 is

$$r_1{}^2 - r_1(b^2 - a^2\cos\theta_1)/b + f^2 = 0 \quad\ldots\ldots\ldots\ldots(3),$$

from which it follows that if the line F_1QP cut the inner and outer portions in Q and P, $F_1Q\,.\,F_1P = F_1F^2$; and that the locus of S, the middle point of PQ, is the elliptic limaçon

$$F_1S = \tfrac{1}{2}(b^2 - a^2\cos\theta_1)/b \quad\ldots\ldots\ldots\ldots\ldots(4).$$

The loop is therefore the inverse of the outer portion with respect to the internal focus.

293. It can also be proved, as in §§ 285 and 286, that the tangents at the extremities of a chord F_1QP drawn through the internal focus intersect at a point T on the circle circumscribing FPQ; that T is the middle point of the arc QFP; and that the locus of T is the cissoid $x(x^2 + y^2) = by^2$.

It can also be shown, as in § 287, that the locus of the intersection of tangents at the extremities of a chord through the nodal focus is a circular cubic.

294. *The tangents at the points P and Q subtend at the node angles which are supplementary.*

Since the points $FTPQ$ are concyclic,
$$TFP = TQP = TPQ = \pi - TFQ.$$

295. *If a chord through the internal focus F_1 meet the loop in Q and Q', the angles which the tangents at Q and Q' make with FQ, FQ' are complementary.*

Let O be the centre of the circle through $FTPQ$; then, since this circle touches BX at F,
$$XFT = FQT = \tfrac{1}{2}FOT = \tfrac{1}{2}QF_1F.$$

Similarly, if the tangent at Q' meet the circle through $FQ'P'$ in T', it follows that
$$XFT' = FQ'T' = \tfrac{1}{2}Q'F_1F,$$
whence
$$XFT + XFT' = FQT + FQ'T'$$
$$= \tfrac{1}{2}(QF_1F + Q'F_1F) = \tfrac{1}{2}\pi,$$

therefore the angles XFT and XFT' and also the angles FQT and $FQ'T'$ are complementary. Whence the angle TFT' is a right angle.

If the chord through F_1 meets the outer portion in P and P', a similar proposition holds good; hence the theorem may be enunciated as follows:—

If a chord through the internal focus cut the curve in four points, and these four points be joined to the node, the angles which the tangents at any two of these points make with their respective radii are equal or complementary according as the two points are or are not the inverse points with respect to the internal focus.

296. *If a chord through the internal focus F_1 cut the outer portion in P, P' and the loop in Q, Q'; and if the tangents at P and Q intersect at T; those at P and Q' in T''; those at P and P' in T_2; and those at Q and Q' in T_1; then the following relations exist between the angle at which any pair of tangents intersect, and the angle which the corresponding chord of contact subtends at the nodal focus, viz.*

$$\text{(i)} \quad PTQ = PFQ.$$

$$\text{(ii)} \quad PT''Q' = PFQ' - \tfrac{1}{2}\pi.$$

$$\text{(iii)} \quad PT_2P' = PFP' - \tfrac{1}{2}\pi.$$

$$\text{(iv)} \quad QT_1Q' = \tfrac{1}{2}\pi - QFQ'.$$

(i) Since the points $FTPQ$ are concyclic, the first proposition at once follows.

(ii) From the figure, we have

$$PT''Q' = Q' - P = Q' - Q.$$

Also
$$PFQ' = PFQ + QFQ'.$$

But
$$PFQ = PTQ = \pi - 2Q,$$

and
$$QFF_1 = FPF_1 = Q - FPT$$
$$= Q - \tfrac{1}{2}O = Q - \tfrac{1}{2}FF_1Q.$$

Similarly
$$Q'FF_1 = Q' - \tfrac{1}{2}\pi + \tfrac{1}{2}FF_1Q,$$

whence
$$QFQ' = Q + Q' - \tfrac{1}{2}\pi.$$

Accordingly
$$PFQ' = Q' - Q + \tfrac{1}{2}\pi$$
$$= PT''Q' + \tfrac{1}{2}\pi.$$

(iii) We have

$$PT_2P' = \pi - P - P'$$
$$= \pi - Q - Q',$$

also since the triangles FF_1Q and FPF_1 are similar

$$PFF_1 = FQF_1 = \pi - Q - FQT$$
$$= \pi - Q - \tfrac{1}{2}O$$
$$= \pi - Q - \tfrac{1}{2}FF_1Q.$$

Similarly
$$P'FF_1 = \tfrac{1}{2}\pi - Q' + \tfrac{1}{2}FF_1Q,$$

whence
$$PFP' = \tfrac{3}{2}\pi - Q - Q' = \tfrac{1}{2}\pi + PT_2P'.$$

(iv)   We have

$$QT_1Q' = \pi - Q - Q'$$
$$= \tfrac{1}{2}\pi - QFQ'$$

by (ii).

The Trisectrix.

297.   One of the most interesting applications of the hyperbolic limaçon is the trisection of an angle. This is effected by means of the trisectrix, which is the name given to the curve when $b = 2a$.

Let α be the angle which is to be trisected. Through the vertex A of the loop draw a line PQR, cutting the trisectrix in P, Q and R such that the angle $PAF = \alpha$. Join FR, and draw AE parallel to FR. Let $RFA = \theta$. Then since $b = 2a$,

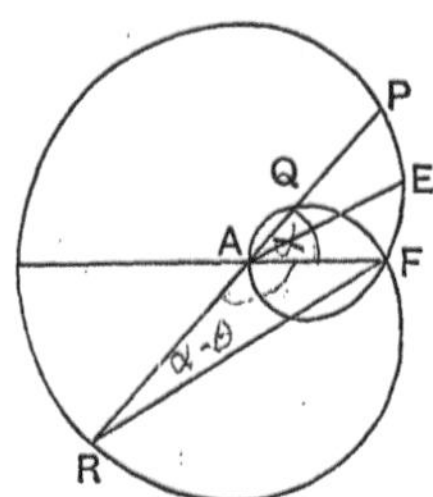

$$FR = a\,(2\cos\theta + 1).$$

But

$$\frac{FR}{a} = \frac{\sin\alpha}{\sin(\alpha - \theta)} = 2\cos\theta + 1,$$

whence

$$\cot^2\alpha = \frac{(2\cos\theta - 1)^2\,(1 + \cos\theta)}{(2\cos\theta + 1)^2\,(1 - \cos\theta)}$$
$$= \frac{1 + \cos 3\theta}{1 - \cos 3\theta};$$

therefore

$$\cos 3\theta = \cos 2\alpha,$$

whence

$$3\theta = 2\alpha.$$

Accordingly

$$EAF = \theta = \tfrac{2}{3}\alpha = \tfrac{2}{3}PAF,$$

whence EA trisects the angle PAF.

The Cardioid.

298.   When $a = b$, the limaçon becomes a cardioid, and its equation may be written in either of the forms

$$\left.\begin{aligned} r &= a\,(1 - \cos\theta) \\ r^{\frac{1}{2}} &= a^{\frac{1}{2}}\sqrt{2}\,.\,\sin\tfrac{1}{2}\theta \end{aligned}\right\}\dots\dots\dots\dots\dots(1),$$

or

$$(x^2 + y^2)^2 + 2ax\,(x^2 + y^2) = a^2 y^2.$$

The cardioid (i) is the inverse of a parabola with respect to its focus, (ii) it is the pedal of a circle with respect to a point on its circumference, (iii) it belongs to the class of curves $r^n = a^n \sin n\theta$, whose properties have been discussed in § 67.

We shall prove in the next section that the cardioid also belongs to the important class of curves called epicycloids, since it is the curve traced out by a point on a circle which rolls outside another circle of equal radius.

The cusp F of a cardioid is a triple focus, since it is the limiting position of the three single foci of an oval of Descartes.

299. *To prove that the cardioid is a one-cusped epicycloid.*

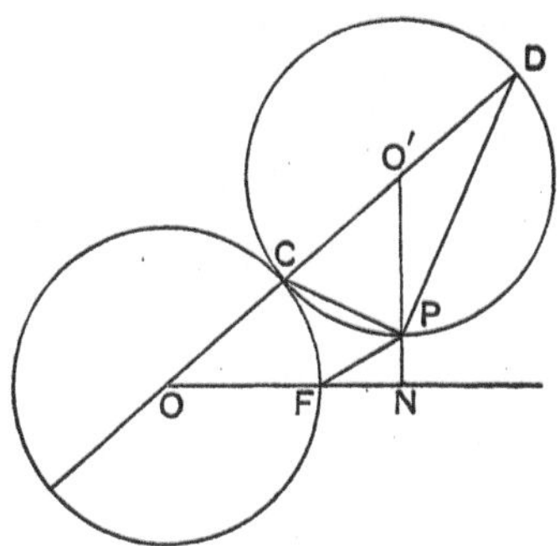

Let P be the point on the rolling circle which initially coincided with F. Then $PFN = FOO' = PO'O = \theta$. Let $OF = \frac{1}{2}a$. Then if F is the origin

$$ON = \tfrac{1}{2}a + r \cos \theta = a \cos \theta - \tfrac{1}{2}a \cos 2\theta,$$

whence the locus of P is

$$r = a\,(1 - \cos \theta).$$

It follows from § 280 that the centre of the fixed circle is the triple focus corresponding to the point of intersection of the two imaginary cuspidal tangents.

It will be shown hereafter that the evolute of an epicycloid is another epicycloid having the same number of cusps; hence the evolute of a cardioid is another cardioid.

300. The line PD is obviously the tangent at P, hence $\phi = FPY = \frac{1}{2}\theta$. Let A be the vertex, F the cusp; PT the tangent

at P, FY perpendicular to PT, $AFY = \chi$. Then the following results can be easily proved :

$$FP^3 = FA \cdot FY^2 \dotfill (2),$$

$$FY = 2a \cos^3 \tfrac{1}{3}\chi \dotfill (3).$$

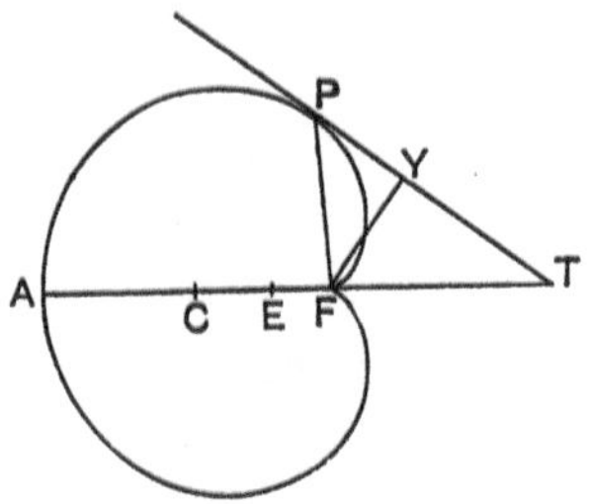

The first equation is the p and r equation of the curve; whilst the second is the pedal with respect to the cusp, or it may be regarded as the tangential polar equation of the curve.

Another form of the tangential polar equation is sometimes useful. Transfer the origin to the triple focus E, then since $FE = \tfrac{1}{2}a$, we obtain

$$p' = FY - \tfrac{1}{2}a \cos \chi$$
$$= \tfrac{1}{2}a \,(4 \cos^3 \tfrac{1}{3}\chi - \cos \chi)$$
$$= \tfrac{3}{2}a \cos \tfrac{1}{3}\chi \dotfill (4).$$

Equations (3) and (4) are the tangential polar equations of the curve referred to the cuspidal focus F and the triple focus E respectively.

From (3) or directly, the tangential equation in Boothian coordinates is

$$27a^2\,(\xi^2 + \eta^2) = 2\,(2 + a\xi)^3 \dotfill (5).$$

Equation (5) shows that the cubic

$$27\,(x^2 + y^2)\,c = 2\,(2c + x)^3$$

is the reciprocal polar of a cardioid; and if the origin be transferred to the point $x = 4c$, the cubic becomes

$$2x^3 = 9c\,(x^2 - 3y^2),$$

which has a crunode at the origin, and therefore two of its three points of inflexion are imaginary. Hence a cardioid has one real double tangent, one real cusp at the origin, and two imaginary cusps at the circular points at infinity. Also the curve is of the third class. The curve accordingly belongs to the tenth division of quartics.

301. *Cuspidal chords are of constant length; and tangents at their extremities intersect at right angles on a fixed circle, whose centre is the triple focus.*

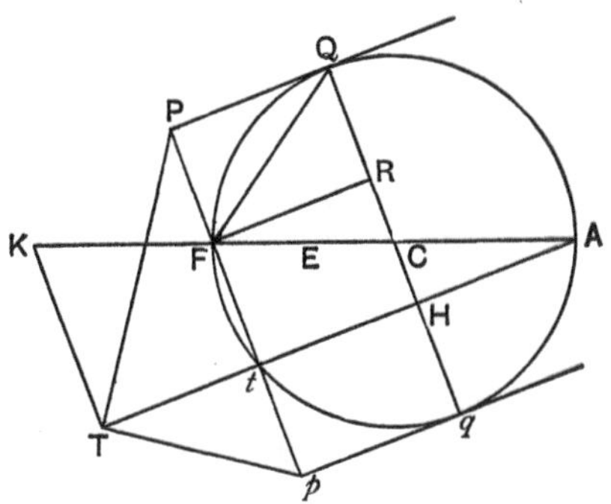

Let F be the cusp, C the centre of the circle of which the cardioid is the pedal; Qq any diameter of the circle. Then if Pp be drawn through F parallel to Qq cutting the tangents at Q and q to the circle in P and p; Pp is a cuspidal chord of the cardioid, and is equal to Qq.

Let the tangents at P, p intersect in T; then

$$TPF = FQP,$$

$$TpP = Fqp = FQq = \tfrac{1}{2}\pi - FQP,$$

whence $\qquad TPp + TpP = \tfrac{1}{2}\pi,$

and therefore PTp is a right angle. Also the triangles QFq and pTP are equal in every respect.

Join At, Tt, where A is the other extremity of the diameter FC; then $tp = FP$, $Tp = FQ$, also $TpP = PFQ$, whence the triangles Tpt and QFP are equal in every respect, and therefore $Ttp = \tfrac{1}{2}\pi$, and TtA is a straight line.

Draw TK perpendicular to AT; then since $Tt = FR = \tfrac{1}{2}At$; $AK = \tfrac{2}{3}AF$, and therefore K is a fixed point. The locus is therefore a circle whose centre E is such that $FE = \tfrac{1}{4}FA$.

302. It follows from § 68, that the orthoptic locus of a cardioid must be a sextic curve; hence the circle we have just found is only part of the locus. We shall now prove that:—

If a chord subtend an angle $\frac{1}{3}\pi$ at the cusp, the tangents at the extremities of the chord intersect at right angles on the limaçon[*]

$$r = \tfrac{3}{4}a\,(\sqrt{3} - 2\cos\theta).$$

Let TP, TQ be the tangents, and let TP intersect the cuspidal tangent at t. Then

$$\tfrac{1}{3}\pi = PFQ = QFt - PFt$$
$$= 2\,(FQT - \pi + FPT),$$

whence $\qquad 2\pi - T = FQT + FPT + PFQ = \tfrac{3}{2}\pi,$

whence $\qquad\qquad\qquad T = \tfrac{1}{2}\pi.$

The Cartesian equation of a cardioid is

$$(x^2 + y^2 + ax)^2 = a^2\,(x^2 + y^2),$$

from which it can be shown that the equation of the tangent at a point whose vectorial angle is θ is

$$x\sin\tfrac{3}{2}\theta - y\cos\tfrac{3}{2}\theta = 2a\sin^3\tfrac{1}{2}\theta\ldots\ldots\ldots\ldots(1).$$

Transfer the origin to the point $-\tfrac{1}{4}a$, and write $b = \tfrac{3}{4}a$, and (1) becomes

$$x\sin\tfrac{3}{2}\theta - y\cos\tfrac{3}{2}\theta = b\sin\tfrac{1}{2}\theta\ldots\ldots\ldots\ldots(2),$$

and the equation of the tangent at the point $\theta + \tfrac{1}{3}\pi$ is

$$x\cos\tfrac{3}{2}\theta + y\sin\tfrac{3}{2}\theta = b\sin(\tfrac{1}{2}\theta + \tfrac{1}{6}\pi)\ldots\ldots\ldots\ldots(3).$$

Let $2\chi = \theta$, and square and add (2) and (3) and we obtain

$$x^2 + y^2 = b^2\,\{\sin^2\chi + \sin^2(\chi + \tfrac{1}{6}\pi)\}$$
$$= \tfrac{1}{2}b^2\,\{2 - \cos 2\chi - \cos(2\chi + \tfrac{1}{3}\pi)\}$$
$$= \tfrac{1}{2}b^2\,(2 - \sqrt{3}\cos z)\ \ldots\ldots\ldots\ldots\ldots(4),$$

where $\qquad\qquad\qquad z = 2\chi + \tfrac{1}{6}\pi.$

Eliminating y between (1) and (2) we get

$$2x = 2b\,\{\sin\chi\sin 3\chi + \sin(\chi + \tfrac{1}{6}\pi)\cos 3\chi\}$$
$$= b\,\{\cos 2\chi - \cos 4\chi + \sin(4\chi + \tfrac{1}{6}\pi) - \sin(2\chi - \tfrac{1}{6}\pi)\}$$
$$= b\,(\sqrt{3}\cos z - \cos 2z)\ldots\ldots\ldots\ldots\ldots\ldots(5).$$

---

[*] Wolstenholme, *Proc. Lond. Math. Soc.* vol. IV. p. 327.

Eliminating $\cos z$ between (4) and (5) we obtain

$$8(x^2 + y^2 - b^2)^2 + 6b^2(x^2 + y^2 - b^2) + 3b^3(2x - b) = 0\ldots\ldots(6).$$

Transfer the origin to the point $x = \tfrac{1}{2}b$, and the equation reduces to

$$4(x^2 + y^2)^2 + b(x^2 + y^2)(8x - 3b) + 4b^2x^2 = 0,$$

or
$$r = \tfrac{1}{2}b(\sqrt{3} - 2\cos\theta).$$

303. *The angular points of a given triangle move round the circumference of a fixed circle; prove that the directrices of the system of parabolas which have a given focus and touch the three sides of the triangle envelope a cardioid.*

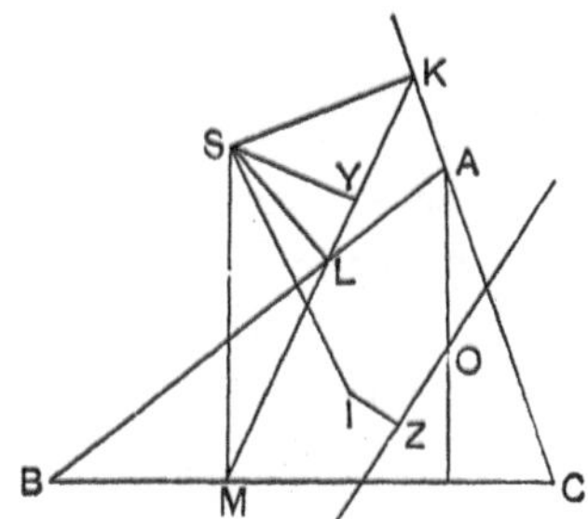

Let ABC be the triangle, O its orthocentre, I the centre of the circle, S the focus of the system of parabolas. Then it is known from the geometry of the parabola (i) that S lies on the circle, (ii) that the directrix of every parabola which touches the sides of the triangle passes through the orthocentre, (iii) the pedal line of S is the tangent at the vertex of the parabola; hence the directrix is parallel to the pedal line KLM.

Draw IZ perpendicular to the directrix, and let $SIZ = \psi$, $MKC = \phi$. Then

$$p = IZ = 2SY + R\cos\psi,$$

and
$$SY = R(\cos B + \cos ISK)\cos\phi\,;$$

also
$$ISK = IAS + ASK$$

$$= IAL + SKL + KLA$$

$$= \pi - 2\phi - C + A,$$

and
$$ISK = \pi - \psi + \phi,$$

whence
$$\psi = 3\phi + C - A.$$

Accordingly

$$p = - 2R \cos (A + C) \cos \phi - 2R \cos (2\phi + C - A) \cos \phi$$
$$+ R \cos (3\phi + C - A)$$
$$= - 2R \cos (A + C) \cos \phi - R \cos (\phi + C - A)$$
$$= - 2R \cos (A + C) \cos \tfrac{1}{3} (\psi - C + A) - R \cos \tfrac{1}{3} (\psi + 2C - 2A)$$
$$\dots\dots\dots(1).$$

Let

$$\kappa \cos \alpha = 2 \cos (A + C) \cos \tfrac{1}{3} (C - A) + \cos \tfrac{2}{3} (C - A),$$
$$\kappa \sin \alpha = 2 \cos (A + C) \sin \tfrac{1}{3} (C - A) - \sin \tfrac{2}{3} (C - A),$$

then

$$\kappa^2 = 1 - 8 \cos A \cos B \cos C = IO^2/R^2.$$

Accordingly (1) becomes

$$p = - IO \cos (\tfrac{1}{3} \psi - \alpha).$$

This is the tangential polar equation of a cardioid referred to the centre of the circle, which is the locus of the points of intersection of tangents at the extremities of cuspidal chords, as origin; and the radius of this circle is equal to IO.

304. *A parabola is described touching a given circle and having its focus at a given point on the circle; prove that the envelope of its directrix is a cardioid.*

Let C be the centre of the circle, S the focus of the parabola, P the point of contact; draw SX perpendicular to the directrix and meeting it in X. Let $XSP = \theta$, $XSC = \psi$, $SC = c$; then

$$\psi - \theta = CSP = CPS = \tfrac{1}{2} \theta,$$

whence $2\psi = 3\theta$. Also

$$SX = SP (1 + \cos \theta) = 2c \cos^3 \tfrac{1}{2} \theta$$
$$= 2c \cos^3 \tfrac{1}{3} \psi,$$

which is the tangential polar equation of a cardioid.

The Conchoid of Nicomedes.

305. This curve was invented by the Greek geometer Nicomedes for the purpose of trisecting an angle, and may be described as follows. Let O be a fixed point and AB a fixed straight line; let $OA = a$, and draw a straight line OQ cutting AB in Q, and

produce it to P so that $PQ = b$; then the locus of P is the conchoid.

Let $AQO = \theta$, $OP = r$, then the polar equation of the curve is

$$r = a \operatorname{cosec} \theta + b \dots\dots\dots\dots\dots\dots (1),$$

or
$$(x^2 + y^2)(y - a)^2 = b^2 y^2 \dots\dots\dots\dots\dots (2).$$

The origin is a double point, the tangents at which are

$$a^2 x^2 + (a^2 - b^2) y^2 = 0,$$

and is therefore a node, a cusp or a conjugate point, according as $a <$ or $=$ or $> b$; also the line $y = a$ is an asymptote.

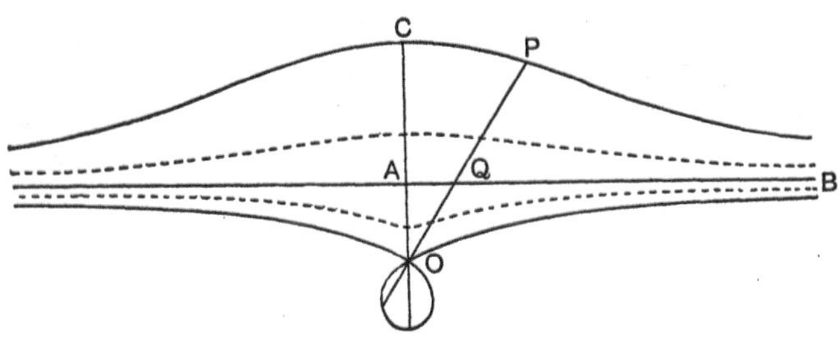

The form of the curve is shown in the figure; the dotted line represents the curve when $a > b$, and the dark line when $a < b$.

The curve has also a real tacnode at infinity on the asymptote; for if the origin be transferred to the point A, (2) may be written in the form

$$- a^2 b^2 - 2aby + y^2 \{x^2 + (y + a)^2 - b^2\} = 0,$$

which is of the same form as the first equation of § 188. If therefore the point O is a node, the curve is a trinodal quartic and belongs to species VII; if on the other hand O is a cusp, the curve belongs to species VIII, and is of the fifth class.

The curve obviously passes through the circular points; hence a circle which passes through the node cannot intersect the conchoid in more than four other points. Also if the equation of the upper portion be $r = a \operatorname{cosec} \theta + b$, that of the lower portion will be $r = a \operatorname{cosec} \theta - b$.

306. We shall now show how the conchoid can be employed to trisect an angle.

Let $POM = \phi$ be the angle which is to be trisected. Bisect

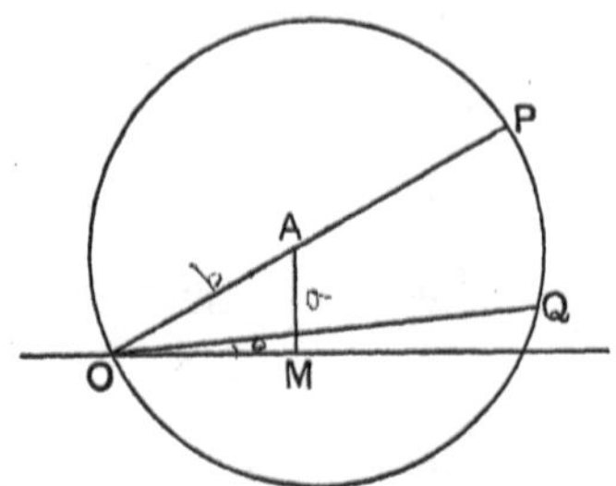

OP in A, and draw AM perpendicular to OM. Let $OA = b$, $AM = a$.

Through P draw the conchoid $r = a \operatorname{cosec} \theta + b$, the origin being O and the initial line OM. With A as a centre describe a circle of radius AO cutting the other branch of the conchoid in Q. Then the angle $QOM = \tfrac{1}{3} POM$.

Let $QOM = \theta$, then since Q lies in the lower branch

$$OQ = a \operatorname{cosec} \theta - b,$$

also
$$AM = a = b \sin \phi,$$

and
$$OQ = 2b \cos (\phi - \theta),$$

whence
$$2 \cos (\phi - \theta) \sin \theta = \sin \phi - \sin \theta,$$

or
$$\sin (\phi - 2\theta) = \sin \theta,$$

and therefore
$$\phi = 3\theta \, ;$$

accordingly OQ trisects the angle POM.

CHAPTER XI.

MISCELLANEOUS CURVES.

307. In the present chapter we shall consider a variety of miscellaneous curves, some of which like the cycloid and catenary are transcendental ones, whilst others like the three- and four-cusped hypocycloids are algebraic curves which are particular cases of a general class of transcendental curves.

The Cycloid.

308. The cycloid is the curve traced out by a point on the circumference of a circle which rolls on a straight line.

To find the equation of the cycloid.

Let a be the radius of the rolling circle CPG, which rolls on the line AG; and let P be the point which initially coincided

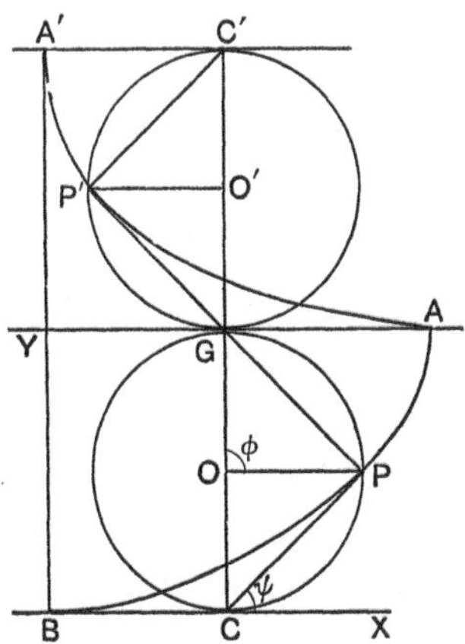

with A. Let (x, y) be the coordinates of P referred to BX an BY, as axes of x and y; and let $GOP = \phi$. Then

$$AG = \text{arc } GP = a\phi,$$

and therefore

$$AY = a\pi.$$

Now

$$\left. \begin{array}{l} x = a\,(\pi - \phi) + a \sin \phi \\ y = a + a \cos \phi \end{array} \right\} \quad \dots\dots\dots\dots\dots(1),$$

whence

$$x = a \cos^{-1} (a - y)/a + (2ay - y^2)^{\frac{1}{2}}.$$

Since PC is the direction of motion of P, PC is the tangen and PG is the normal at P.

309. *The evolute of a cycloid is an equal cycloid.*

Let $PCX = \psi$; then $\phi = \pi - 2\psi$, and (1) becomes

$$\left. \begin{array}{l} x = 2a\psi + a \sin 2\psi \\ y = a - a \cos 2\psi \end{array} \right\} \quad \dots\dots\dots\dots\dots(2),$$

whence

$$\frac{ds}{d\psi} = \rho = 4a \cos \psi \quad \dots\dots\dots\dots\dots(3),$$

and

$$s = 4a \sin \psi \quad \dots\dots\dots\dots\dots(4),$$

no constant being required, since $s = 0$ when $\psi = 0$.

Equation (3) shows that $\rho = 2PG$, whence if P' be the centre of curvature of P, the evolute is another equal cycloid $A'P'A$, whose vertex A coincides with the cusp of the original cycloid.

Equation (4) proves the isochronism of the cycloid; for the equation of motion of a particle sliding down a cycloidal tube under the action of gravity is

$$\frac{d^2s}{dt^2} + \mu g \sin \psi = 0,$$

or

$$\frac{d^2s}{dt^2} + (\mu g/4a)\,s = 0,$$

whence the time of motion from any point P to B is $\pi\,(a/\mu g)^{\frac{1}{2}}$.

Squaring and adding (3) and (4) we get

$$BP^2 + PP'^2 = CC'^2.$$

310. *If a parabola be described which touches a cycloid at the vertex B, and whose latus-rectum is the line joining the adjacent pair of cusps, any double ordinate to the parabola drawn from a point on the arc joining the extremities of the latus-rectum is equal to the intercepted arc of the cycloid.*

Through P draw a line PM perpendicular to BY, and cutting the parabola in Q. Then

$$QM^2 = 8a \cdot BM = 16a^2 \sin^2 \psi$$

by the second of (2); whence

$$QM = 4a \sin \psi = BP.$$

311. *To find the tangential equation of the cycloid.*

Let PC meet YB produced in R; then

$$\xi^{-1} = BC = 2a\psi,$$
$$\tan \psi = -\xi/\eta,$$

whence
$$1 + 2a\xi \tan^{-1} \xi/\eta = 0 \quad \dots\dots\dots\dots\dots(5).$$

Epicycloids and Hypocycloids.

312. The epicycloid is the curve traced out by a point on the circumference of a circle which rolls outside another circle.

To find the equation of an epicycloid.

Let a and b be the radii of the fixed and rolling circles, so that $OQ = a$, $O'Q = b$; also let* $QOA = \theta$, $PO'Q = \phi$, where P is the point which initially coincided with A. Then since

$$\text{arc } AQ = \text{arc } PQ, \quad a\theta = b\phi;$$

whence the coordinates of P are given by the equations

$$\begin{aligned}
x &= (a+b) \cos\theta - b \cos(a+b)\theta/b \\
y &= (a+b) \sin\theta - b \sin(a+b)\theta/b
\end{aligned} \quad \dots\dots\dots(6);$$

the elimination of θ between these equations determines the Cartesian equation of the curve.

The line EP is the tangent to the curve at P, whence if $OY = p$, where OY is perpendicular to EP, we have

$$OTE = \pi - \psi, \quad p = (a+2b) \sin \tfrac{1}{2}\phi;$$

also
$$\psi = \theta + \tfrac{1}{2}\phi = \tfrac{1}{2}(a+2b)\phi/a,$$

whence
$$p = (a+2b) \sin \frac{a\psi}{a+2b} \dots\dots\dots\dots\dots(7).$$

---

* The point A (not marked in the figure), is the point between O and T where the fixed circle cuts OT.

This is the tangential polar equation of an epicycloid and is of the form $p = c \sin n\theta$; it is also the pedal of the curve with respect to the centre of the fixed circle, and the inverse curve is the reciprocal polar of the epicycloid.

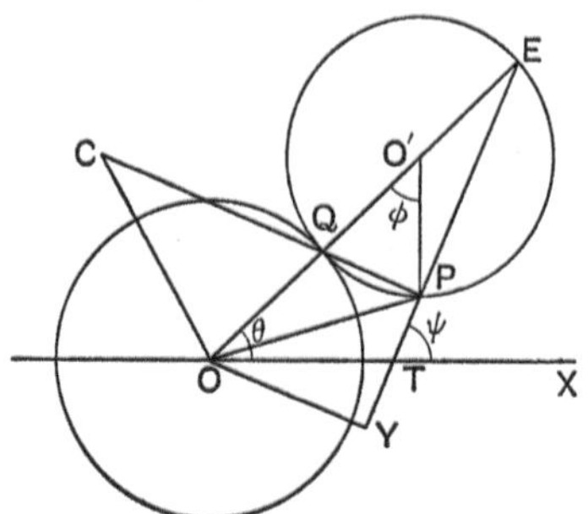

Again
$$\rho = p + \frac{d^2 p}{d\psi^2} = \frac{4b\,(a+b)}{(a+2b)^2}\,p \dots\dots\dots\dots(8),$$

which shows that the radius of curvature is proportional to the perpendicular from the centre of the fixed circle on to the tangent. Also since $\rho = ds/d\psi$, we obtain from (7) and (8)

$$s = \frac{4b\,(a+b)}{a}\left(1 - \cos\frac{a\psi}{a+2b}\right) \dots\dots\dots\dots(9),$$

which is the intrinsic equation of the curve, s being measured from A.

The p and r equation of the curve seems to have been first given by the Jesuits in their notes to Prop. LI. of Newton's *Principia*, and may be obtained as follows. Let $OP = r$; then

$$r^2 = (a+b)^2 + b^2 - 2\,(a+b)\,b\cos\phi$$
$$= a^2 + 4\,(a+b)\,b\sin^2 \tfrac{1}{2}\phi$$
$$= a^2 + \frac{4\,(a+b)\,bp^2}{(a+2b)^2},$$

whence
$$p^2 = \frac{(a+2b)^2}{4\,(a+b)\,b}\,(r^2 - a^2) \dots\dots\dots\dots\dots(10).$$

313.   *The evolute of an epicycloid is a similar epicycloid.*

The evolute of the curve is the envelope of the normal PQ. Now if OZ be the perpendicular from O on to the normal

$$OZ = a\cos\tfrac{1}{2}\phi = a\cos a\psi/(a+2b),$$

which is the tangential polar equation of a similar epicycloid.

All cycloidal curves belong to the class of curves called *roulettes,* for the complete discussion of which we must refer to Dr Besant's *Notes on Roulettes and Glissettes.* We shall only give one proposition on the subject.

314. *A curve rolls on a straight line; it is required to find the roulette traced by any point Q.*

Let QA be a line fixed in the plane of the rolling curve AP,

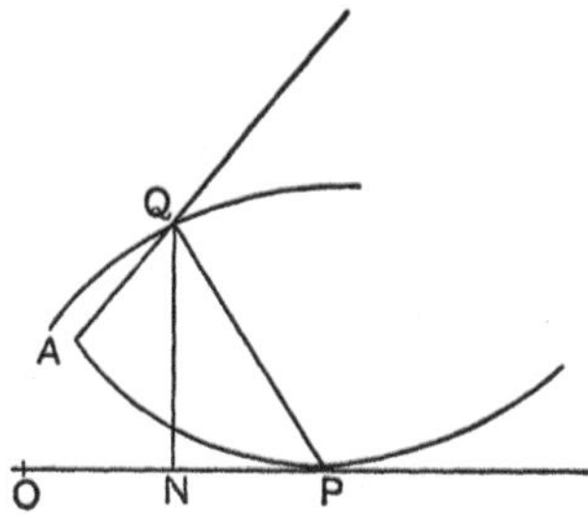

and let OP be the line on which it rolls. Let A initially coincide with O. Then if (x, y) be the coordinates of Q referred to O,

$$ON = x, \quad QN = y;$$

also if (r, θ) be the polar coordinates of P referred to QA, in the plane of the rolling curve

$$AQP = \theta, \quad QN = p, \quad QP = r, \quad QPN = \phi,$$

whence
$$\left. \begin{array}{l} y = p, \quad \tan \phi = dx/dy = rd\theta/dr \\ y = r \sin \phi = rdx/ds \end{array} \right\} \quad \dots\dots\dots\dots(11).$$

315. *If the roulette is an ellipse, the rolling curve is an epicycloid.*

Let the roulette be
$$x^2/a^2 + y^2/b^2 = 1,$$

then
$$\left(\frac{dx}{dy}\right)^2 = \frac{a^2 y^2}{b^2 (b^2 - y^2)},$$

whence by (11)
$$\frac{a^2 p^2}{b^2 (b^2 - p^2)} = \tan^2 \phi = \frac{p^2}{r^2 - p^2},$$

or
$$r^2 = b^4/a^2 + e^2 p^2,$$

which is the p and r equation of an epicycloid. If therefore an epicycloid roll on a straight line, the locus of the centre of the fixed circle is an ellipse.

316. *To find the tangential equation of an epicycloid.*

Through O draw a line OY' perpendicular to OT cutting the tangent in Y'; then

$$\frac{OT}{OE} = \frac{\sin \frac{1}{2}\phi}{\sin (\frac{1}{2}\phi + \theta)},$$

whence

$$\xi = \frac{\sin (\frac{1}{2}a/b + 1)\theta}{(a + 2b)\sin \frac{1}{2}a\theta/b} \dots\dots\dots\dots\dots(12);$$

$$\frac{OY'}{OE} = \frac{\sin \frac{1}{2}\phi}{\cos (\frac{1}{2}\phi + \theta)},$$

whence

$$\eta = \frac{\cos (\frac{1}{2}a/b + 1)\theta}{(a + 2b)\sin \frac{1}{2}a\theta/b} \dots\dots\dots\dots\dots(13);$$

the elimination of θ between these equations furnishes the required result.

The one-cusped epicycloid, as we have already shown, is the cardioid, and its tangential equation is

$$27a^2 (\xi^2 + \eta^2)(1 - a\xi) = 4 \quad \dots\dots\dots\dots(14),$$

whilst those of the two- and four-cusped epicycloids are

$$4a^2 (\xi^2 + \eta^2)(1 - a^2\xi^2) = 1 \dots\dots\dots\dots\dots(15)$$

and

$$(6b)^6 (\xi^2 + \eta^2)\,\xi^2\eta^2 = 4 \{27b^2 (\xi^2 + \eta^2) - 1\}^2\dots\dots\dots(16)$$

respectively. The former curve is of the fourth class and it cannot be of lower degree than the sixth, since the common tangent at the two cusps has a contact of the second order with the curve at these points. Its Cartesian equation is

$$4 (x^2 + y^2 - a^2)^3 = 27a^4y^2 \dots\dots\dots\dots\dots(17).$$

317. The hypocycloid is the curve traced out by a point on the circumference of a circle which rolls inside a fixed circle.

The coordinates of any point on a hypocycloid can be shown to be

$$\left.\begin{array}{l} x = (a - b)\cos \theta + b \cos (a - b)\,\theta/b \\ y = (a - b)\sin \theta - b \sin (a - b)\,\theta/b \end{array}\right\} \quad \dots\dots(18),$$

whence the corresponding results for a hypocycloid can be deduced from those of an epicycloid by changing the sign of b.

A two-cusped hypocycloid is a diameter of the fixed circle, as can be at once shown by elementary geometry. The three-

and four-cusped hypocycloids possess many remarkable properties, and will be discussed separately.

318.   In the preceding section we have tacitly supposed that the radius of the fixed circle is greater than that of the moving circle; we shall now show that if the radius of the latter is greater than that of the former, the hypocycloid becomes an epicycloid generated by a rolling circle whose radius is equal to the difference between the radii of the two circles.

Let AQB and QPR be the fixed and rolling circles, O, O_1 their centres; a and b their radii; and let E be the point with which

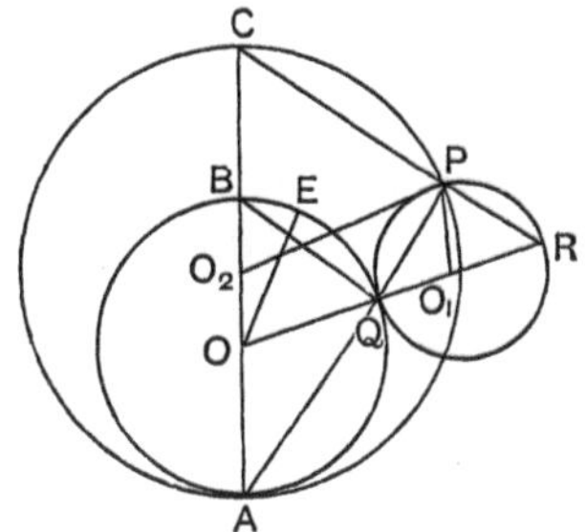

P was initially in contact.   Let

$$EOQ = \theta, \quad QO_1P = \phi.$$

Let R be the other extremity of the diameter through Q of the moving circle, and draw RP to meet the diameter AB of the fixed circle in C.

Since BQA and CPA are right angles, and $OQ = OB$, it follows that $CB = QR$; whence

$$AC = 2a + BC = 2(a + b);$$

whence C is a fixed point, and a semicircle can be described through APC whose radius is $a + b$, and centre O_2.

Again   arc $AE = a(\theta + \phi) = (a + b)\phi$,

arc $AP = (a + b)AO_2P = (a + b)\phi$,

whence arc $AP = $ arc AE.   Accordingly the epicycloid which is the locus of P may be generated by the circle APC whose radius is $a + b$ rolling on the fixed circle in such a manner that their concavities are in the same direction.

14—2

Trochoidal Curves.

319. The trochoid is a curve described by any point in the plane of a circle which rolls on a straight line.

Trochoids are sometimes called prolate cycloids and curtate cycloids according as the point lies within or without the rolling circle. The equations of a trochoid can easily be shown to be

$$\left.\begin{array}{l} x = a\,(\theta - m\sin\theta) \\ y = a\,(1 - m\cos\theta) \end{array}\right\} \quad \dots\dots\dots\dots\dots(19),$$

where m is less or greater than unity according as the point lies within or without the circle.

320. When a circle rolls on another circle, the locus of any point in the plane of the moving circle is called an *epitrochoid* or a *hypotrochoid* according as the latter rolls on the exterior or the interior of the fixed circle.

If c be the distance of the point from the centre of the rolling circle, the equations of the epitrochoid are

$$\left.\begin{array}{l} x = (a+b)\cos\theta - c\cos(a+b)\,\theta/b \\ y = (a+b)\sin\theta - c\sin(a+b)\,\theta/b \end{array}\right\} \quad \dots\dots\dots(20),$$

whilst those of the hypotrochoid are obtained by changing the sign of b.

When $a = b$, the epitrochoid becomes a limaçon.

Transfer the origin to the point $x = -c$; then (20) become

$$x = 2\,(a - c\cos\theta)\cos\theta,$$

$$y = 2\,(a - c\cos\theta)\sin\theta,$$

whence θ is the vectorial angle, and the polar equation is

$$r = 2\,(a - c\cos\theta),$$

which is a limaçon.

In (20), put

$$a = (m-1)\,b, \quad mb = c \dots\dots\dots\dots\dots(21)$$

and change the direction of the axis of y, and we obtain

$$x = 2c\sin\tfrac{1}{2}(m+1)\theta\,\sin\tfrac{1}{2}(m-1)\theta,$$

$$y = 2c\cos\tfrac{1}{2}(m+1)\theta\,\sin\tfrac{1}{2}(m-1)\theta;$$

also if ϕ is the angle which the radius vector makes with the axis of y,

$$\phi = \tfrac{1}{2}(m + 1)\,\theta,$$

and the polar equation of the curve becomes

$$r = 2c \sin \frac{m - 1}{m + 1}\,\phi,$$

or

$$r = 2c \sin n\phi \quad \dots\dots\dots\dots\dots\dots\dots(22),$$

where

$$n = (m - 1)/(m + 1).$$

In the case of the epitrochoid, a and b are both positive; accordingly from (21) it follows that $m > 1$, and therefore $0 < n < 1$.

In the case of the hypotrochoid we must put

$$a = (m + 1)\,b, \quad c = mb \quad \dots\dots\dots\dots\dots(23),$$

from which it follows that $m > -1$; and therefore n may have any value which does not lie between zero and unity.

321. *The pedal of an epicycloid or a hypocycloid with respect to the centre of the fixed circle is an epitrochoid or a hypotrochoid.*

For if in (7) we put $a = (m - 1)\,b$, the equation of the pedal is of the form (22). Also the reciprocal polar of an epicycloid or a hypocycloid is the curve

$$r \sin n\theta = c \quad \dots\dots\dots\dots\dots\dots\dots(24).$$

An example of this proposition has already occurred in the case of the cardioid, whose pedal with respect to one of the triple foci is $r = c \cos \tfrac{1}{3}\theta$.

322. *When the radius of the fixed circle is double that of the rolling circle, the hypotrochoid becomes an ellipse.*

Putting $-b$ for b in (20), and then writing $a = 2b$, the equation of the curve becomes

$$\frac{x^2}{(b - c)^2} + \frac{y^2}{(b + c)^2} = 1.$$

323. *To find the orthoptic locus of an epicycloid*[*].

Let $m = a/(a + 2b)$; then by (7) the equation of PY, see figure to § 312, is

$$x \sin \psi - y \cos \psi = (a + 2b) \sin m\psi,$$

* Wolstenholme, *Proc. Lond. Math. Soc.* vol. IV. p. 330.

and that of the perpendicular tangent is

$$x \cos \psi + y \sin \psi = (a + 2b) \sin m \left(\tfrac{1}{2}\pi + \psi\right),$$

whence

$$x = (a + 2b) \left\{\sin m \left(\tfrac{1}{2}\pi + \psi\right) \cos \psi + \sin m\psi \sin \psi\right\},$$
$$y = (a + 2b) \left\{\sin m \left(\tfrac{1}{2}\pi + \psi\right) \sin \psi - \sin m\psi \cos \psi\right\}.$$

These equations may be written in the form

$$x = (a + 2b) \left[\sin \tfrac{1}{4}(1 + m)\,\pi \cos (1 - m)\left(\psi + \tfrac{1}{4}\pi\right)\right.$$
$$\left. + \cos \tfrac{1}{4}(1 + m)\,\pi \sin \left\{(1 + m)\,\psi - \tfrac{1}{4}(1 - m)\,\pi\right\}\right],$$
$$y = (a + 2b) \left[\sin \tfrac{1}{4}(1 + m)\,\pi \sin (1 - m)\left(\psi + \tfrac{1}{4}\pi\right)\right.$$
$$\left. - \cos \tfrac{1}{4}(1 + m)\,\pi \cos \left\{(1 + m)\,\psi - \tfrac{1}{4}(1 - m)\,\pi\right\}\right].$$

Let

$$(1 - m)\left(\psi + \tfrac{1}{4}\pi\right) = \theta,$$

and the equations may be written in the form

$$x = (a + 2b) \left\{\sin \tfrac{1}{4}(1 + m)\,\pi \cos \theta - \cos \tfrac{1}{4}(1 + m)\,\pi \cos \frac{(1 + m)\,\theta}{1 - m}\right\},$$

$$y = (a + 2b) \left\{\sin \tfrac{1}{4}(1 + m)\,\pi \sin \theta - \cos \tfrac{1}{4}(1 + m)\,\pi \sin \frac{(1 + m)\,\theta}{1 - m}\right\},$$

which are the equations of an epitrochoid. Whence if A and B be the radii of the fixed and rolling circles, and C the distance of the fixed point,

$$A + B = (a + 2b) \sin \tfrac{1}{4}(1 + m)\,\pi$$

$$= (a + 2b) \sin \frac{(a + b)\,\pi}{2\,(a + 2b)},$$

$$C = (a + 2b) \cos \frac{(a + b)\,\pi}{2\,(a + 2b)},$$

$$\frac{A + B}{B} = \frac{1 + m}{1 - m} = \frac{a + b}{b}.$$

To verify this result in the case of a cardioid, put $a = b$, and we get

$$A = B = 3\sqrt{3}a/4, \quad C = 3a/2,$$

and the locus is the limaçon

$$r = \tfrac{3}{2}a\left(\sqrt{3} - 2 \cos \theta\right),$$

which agrees with § 302.

The Three-cusped Hypocycloid.

324. The three-cusped hypocycloid is the curve traced out by a point on the circumference of a circle which rolls inside another circle of three times its radius.

Putting $a = 3b$ in (18) of § 317 we obtain

$$\left. \begin{array}{l} x/b = 2 \cos \theta + \cos 2\theta \\ y/b = 2 \sin \theta - \sin 2\theta \end{array} \right\} \quad \dots\dots\dots\dots\dots(1),$$

whence squaring and adding we get

$$\tfrac{1}{4}(r^2 - 5b^2) = b^2 \cos 3\theta = b^2 \cos \theta (4 \cos^2 \theta - 3).$$

But from the first of (1) we get

$$x + b = 2b \cos \theta (1 + \cos \theta) \dots\dots\dots\dots\dots(2),$$

whence
$$\frac{r^2 - 5b^2}{2b(x+b)} = \frac{4\cos^2 \theta - 3}{1 + \cos \theta} \quad \dots\dots\dots\dots\dots(3).$$

Substituting the value of $\cos^2 \theta$ from (2) in (3) we get

$$(1 + \cos \theta)(r^2 + 8bx + 3b^2) = 2(x + b)(2x + 3b)\dots\dots(4).$$

Eliminating θ between (2) and (4) we obtain

$$(r^2 + 12bx + 9b^2)^2 = 4b(2x + 3b)^3,$$

or
$$r^4 + 18b^2r^2 - 8bx^3 + 24bxy^2 = 27b^4\dots\dots\dots\dots(5),$$

which shows that the curve is a *tricuspidal quartic*; and therefore belongs to species X.

The tangential equation of the curve is

$$\xi^2 + \eta^2 = b\xi(3\eta^2 - \xi^2),$$

which shows that the only double tangent the curve has is the line at infinity, which touches the curve at the circular points.

325. *The orthoptic locus of a three-cusped hypocycloid is a circle.*

Let O and O' be the centres of the fixed and moving circles, and let A be the initial position of the moving point P. Then if OO' cuts the moving circle in E and Q, PE is the tangent at P. Let TP' be the perpendicular tangent.

Then by the mode of generation

$$PO'Q = 3AOQ = 2PEQ;$$

if therefore the tangent PE meet OA in t,

$$AOQ = 2OtE\dots\dots\dots\dots\dots\dots\dots\dots\dots\dots\dots(6).$$

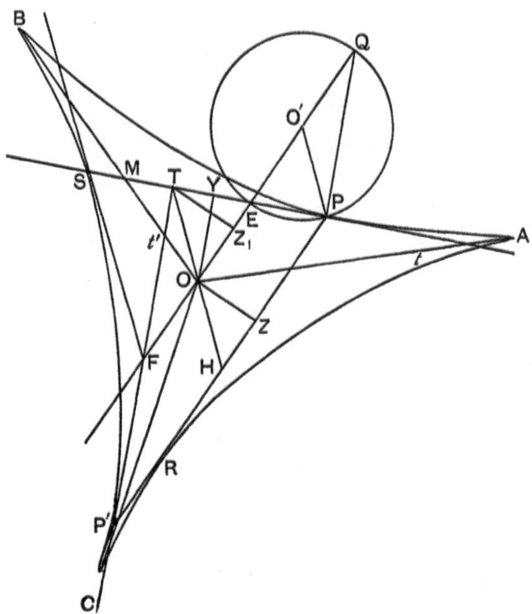

Now if TP' cut OB in t',

$$t'OE = BOA - AOQ$$
$$= \tfrac{2}{3}\pi - 2OtE.$$

Also by (6) if OF be drawn through the centre of the moving circle corresponding to P',

$$t'OF = 2Ot'F$$
$$= 2\left(\tfrac{1}{2}\pi - OMA\right)$$
$$= \tfrac{1}{3}\pi + 2OtE,$$

whence
$$t'OE + t'OF = \pi,$$

and therefore EO and OF are in the same straight line; accordingly

$$OE = OF = OT,$$

and therefore the locus of T is a circle whose radius is equal to that of the moving circle.

326. *To find the tangential polar and the intrinsic equations of the curve.*

Let
$$OY = p, \quad OE = a, \quad OtY = \theta,$$
then since we have shown that $OEY = 3\theta$, it follows that
$$p = a \sin 3\theta,$$
or
$$p = -a \cos 3\phi,$$
if $\phi = \tfrac{1}{2}\pi - \theta$. The intrinsic equation is
$$s = \tfrac{8}{3} a (1 - \cos 3\theta),$$
and the p and r equation is
$$r^2 = 8p^2 - 9a^2.$$

327. *The portion of the tangent contained within the curve is of constant length.*

Join PP' and draw OZ perpendicular to PP'. Since $O'PE$ and OET are a pair of equal isosceles triangles, $PE = ET$; similarly $TF = FP'$, and therefore EF is parallel to PP'. Whence $PP' = 2EF = 4OE$.

Produce TO to meet PP' in H; then since O is the middle point of TH, it follows that $OZ = OY$; hence PP' touches the hypocycloid.

The line OH bisects PP' in H; whence if two tangents be drawn to a three-cusped hypocycloid which are at right angles to one another, the chord of contact is also a tangent to the curve, and is bisected by the line joining the centre O with the point of intersection of the perpendicular tangents.

328. *If three tangents be drawn to a three-cusped hypocycloid, two of which are at right angles, the third tangent is perpendicular to the chord of contact of the other two.*

Draw TZ' perpendicular to PP' cutting OE in Z'; let
$$EOA = \phi, \quad EtO = \theta.$$
Then since $OZ'T$ is a right angle
$$OZ' = OT \cos TOZ'.$$

But $OT = a$, and $TOZ' = \pi - 2TEO = \pi - 6\theta = \pi - 3\phi$;
whence
$$OZ' = -a \cos 3\phi,$$
and therefore TZ' touches the hypocycloid.

329. *The locus of the point of intersection of two tangents which intersect at an angle $\frac{2}{3}\pi$ is a curve similar to the pedal of the hypocycloid.*

Let Qt, Qt' be the two tangents which cut OA, OB in t and t'. Draw OY, OZ perpendicular to Qt, Qt', and let

$$QOA = \theta, \quad QtO = \phi, \quad OQ = r.$$

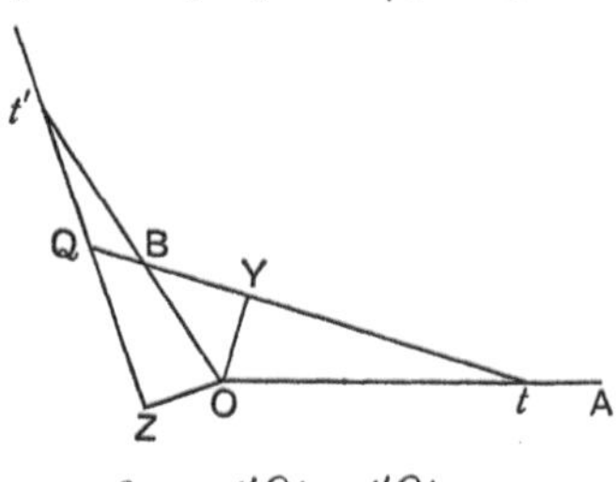

Since $$\tfrac{2}{3}\pi = t'Ot = t'Qt,$$

a circle can be described through $tOQt'$, whence $\phi = QtO = Qt'O$. Accordingly the tangents Qt, Qt' are equally inclined to OA, OB, and therefore the perpendiculars OY, OZ are equal; whence $OQY = \tfrac{1}{6}\pi$, and

$$r = 2OY = 2a \sin 3\phi,$$

and $$\theta + \phi = \tfrac{5}{6}\pi,$$

whence $$r = 2a \cos 3\theta.$$

330. *The envelope of the pedal line of a triangle is a three-cusped hypocycloid, whose centre is the centre of the nine-point circle of the triangle.*

Let P be any point on the circumscribing circle of the triangle ABC; KLM the corresponding pedal line; F the middle point* of

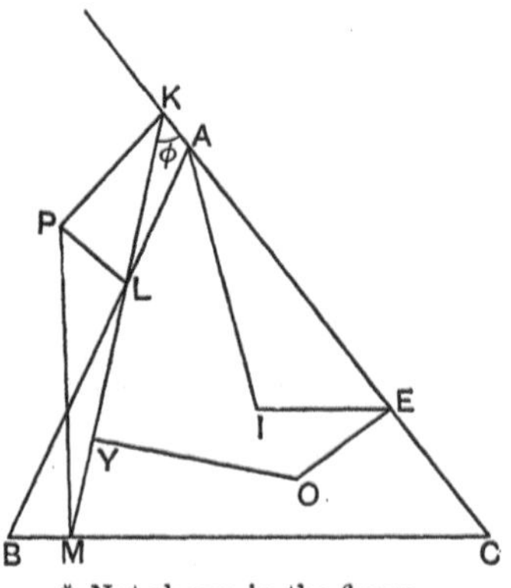

* Not shown in the figure.

AC; I and O the centres of the circumscribing and nine-point circles of the triangle. Draw OY perpendicular to KM, and let

$$OY = p, \quad IA = R, \quad YKA = \phi.$$

Then since
$$IFA = \tfrac{1}{2}\pi,$$

$$p = FK \sin \phi - OF \cos (\phi + OFI)$$
$$= R \sin IPK \sin \phi - \tfrac{1}{2} R \cos (\phi + C - A).$$

Now
$$IPK = IPA + APK = IAP + APK$$
$$= IAL + PKL + KLA$$
$$= \pi - 2\phi - C + A,$$

whence $2p = 2R \sin (2\phi + C - A) \sin \phi - R \cos (\phi + C - A)$

$$= - R \cos (3\phi + C - A),$$

which is the tangential polar equation of a three-cusped hypocycloid, the radius of whose rolling circle is equal to $\tfrac{1}{2}R$.

This theorem seems to have been first discovered by Steiner, and has been discussed by several British mathematicians. The preceding proof is due to Dr Besant.

331. *If a rectangular hyperbola circumscribe a triangle, the envelope of its asymptotes is a three-cusped hypocycloid.*

Let O be the centre of the hyperbola, N that of the nine-point circle of the triangle ABC; E and F the middle points of AC,

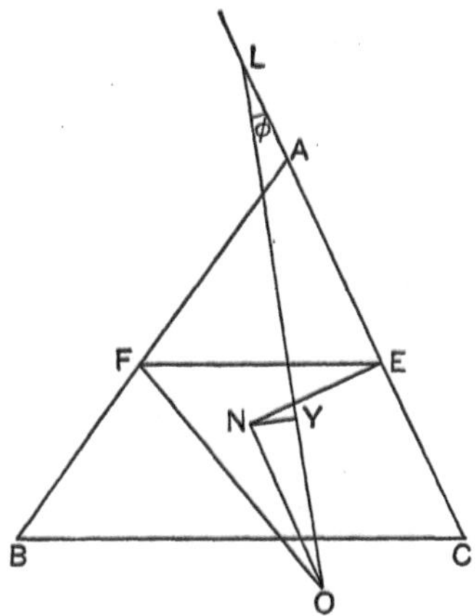

AB. Then O lies in the nine-point circle, and if a point L be taken on AC such that $EL = OE$, OL is an asymptote of the hyperbola.

Draw NY perpendicular to OL, and let $NY = p$, $OLE = \phi$;

then
$$p = \tfrac{1}{2} R \cos ONY,$$

and
$$ONY = \tfrac{1}{2}\pi - NOE + \phi = OFE + \phi\,;$$

also since
$$FOE = A,$$

$$OFE = \pi - A - OEF$$
$$= C - A + 2\phi,$$

whence
$$ONY = C - A + 3\phi,$$

and therefore
$$p = \tfrac{1}{2} R \cos (3\phi + C - A).$$

If a triangle be self-conjugate to a rectangular hyperbola, it is known that the centre of the hyperbola lies on the circle circumscribing the triangle, and that the curve passes through the centres of the inscribed and the three escribed circles of the triangle. Hence the preceding theorem shows that the envelope of the asymptotes of all rectangular hyperbolas to which a given triangle is self-conjugate is a three-cusped hypocycloid, whose centre is the centre of the circumscribing circle of the triangle.

The Four-cusped Hypocycloid.

332. Putting $a = 4b$ in (18) of § 317, the equations of the curve become

$$\left. \begin{aligned} x &= 3b \cos \theta + b \cos 3\theta = a \cos^3 \theta \\ y &= 3b \sin \theta - b \sin 3\theta = a \sin^3 \theta \end{aligned} \right\} \quad \dots\dots\dots\dots(1),$$

whence the Cartesian equation is of the form

$$x^{\frac{2}{3}} + y^{\frac{2}{3}} = a^{\frac{2}{3}} \quad \dots\dots\dots\dots\dots\dots\dots(2).$$

The curve therefore belongs to the class of curves discussed in § 55. Equation (1) may also be expressed in the form

$$(a^2 - x^2 - y^2)^3 = 27 a^2 x^2 y^2 \quad \dots\dots\dots\dots\dots(3),$$

whilst its reciprocal polar is

$$\frac{1}{x^2} + \frac{1}{y^2} = \frac{1}{c^2} \quad \dots\dots\dots\dots\dots\dots(4),$$

which shows that the hypocycloid is of the sixth degree and fourth class.

The characteristics of the curve are most easily investigated by means of the reciprocal curve.

This curve has a complex biflecnode at the origin, and from § 188 it has a pair of real biflecnodes at infinity which are situated on the axes of x and y respectively. The quartic therefore belongs to the seventh species; if therefore we employ the letters δ, κ, τ, ι to denote simple singularities, and the symbols $\delta\iota_2$, $\tau\kappa_2$ to denote a biflecnode and its reciprocal singularity, which by § 166 consists of a pair of cusps having a common cuspidal tangent, Plücker's numbers for the reciprocal polar of a four-cusped hypocycloid are

$$n = 4, \ \delta = 0, \ \kappa = 0, \ m = 6, \ \tau = 4, \ \iota = 0, \ \delta\iota_2 = 3,$$

and therefore the characteristics of the four-cusped hypocycloid are

$$n = 6, \ \delta = 4, \ \kappa = 0, \ m = 4, \ \tau = 0, \ \iota = 0, \ \tau\kappa_2 = 3.$$

All the four nodes are imaginary and are situated on the lines $x \pm y = 0$; whilst two of the singularities $\tau\kappa_2$ are real and consist of the two pairs of cusps on the axes of x and y respectively, and their common cuspidal tangents. The third singularity $\tau\kappa_2$ consists of a pair of imaginary cusps at the circular points and their common cuspidal tangent, which is the line at infinity. This may be proved by writing (3) in the form

$$16\,(a^2 I^2 - \beta\gamma)^3 + 27 a^2 I^2 (\beta^2 - \gamma^2)^2 = 0.$$

The reader will observe that when dealing with sextic and other curves having compound singularities, the latter must be considered *as a whole*, instead of the simple singularities of which they are composed.

333. *The portion of the tangent to a four-cusped hypocycloid, which is intercepted by two real cuspidal tangents, is of constant length.*

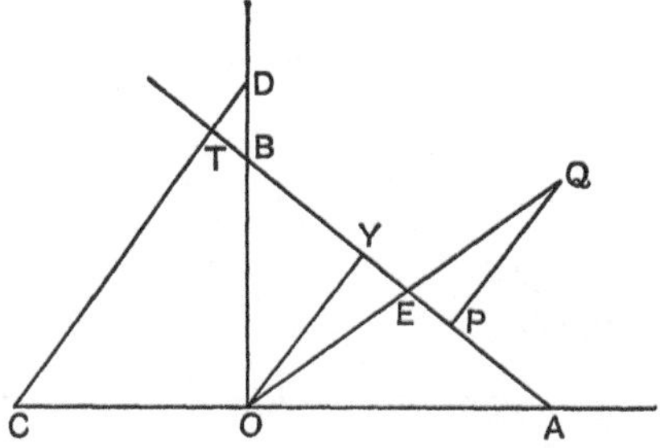

Let AB be a line of constant length a, which slides between two lines at right angles. Through E the middle point of AB

draw OEQ so that $EQ = OE$. Draw OY, QP perpendicular to AB; let $AOY = \phi$, (x, y) the coordinates of P.

Since QA, QB are perpendicular to OA, OB; PB is the direction of motion of P, and therefore AB touches the curve enveloped by it at P. Now

$$x = BP \sin \phi = AY \sin \phi$$
$$= a \sin^3 \phi,$$

similarly $\qquad\qquad y = a \cos^3 \phi,$

whence the locus of P is the hypocycloid

$$x^{\frac{2}{3}} + y^{\frac{2}{3}} = a^{\frac{2}{3}}.$$

Accordingly the four-cusped hypocycloid is the envelope of a straight line of constant length, which slides between two straight lines at right angles to one another.

The equation of the pedal and the p and r equation are respectively

$$r = \tfrac{1}{2} a \sin 2\phi,$$
$$r^2 = a^2 - 3p^2,$$

whence $\qquad\qquad \rho = - r dr/dp = 3p,$

and the intrinsic equation is

$$s = \tfrac{3}{2} a \sin^2 \psi,$$

where $\qquad\qquad \psi = \tfrac{1}{2}\pi - \phi.$

334. *The orthoptic locus is a curve similar to the pedal.*

Let AB and CD be perpendicular tangents intersecting at T; let $OT = r$, $TOA = \theta$. Then since $CD = a$, $OCD = \phi$,

$$TY = a \cos \phi \sin \phi = OY,$$

whence $\qquad\qquad \theta = \tfrac{1}{4}\pi + \phi, \qquad r = p \sqrt{2},$

accordingly $\qquad\qquad r = - (a/2^{\frac{1}{2}}) \cos 2\theta.$

335. *If the tangent at P is the normal at T, then $OP = OT$.*

From the p and r equation we have

$$OT^2 = a^2 - 3TY^2 = a^2 - 3p^2,$$

whence $\qquad\qquad OT = OP.$

The Evolute of an Ellipse.

336. Let (x, y) be the coordinates of the centre of curvature of a point on an ellipse whose excentric angle is ϕ; and let ψ be the angle which the normal makes with the major axis. Then

$$x = a \cos \phi - \rho \cos \psi,$$

and
$$a \cos \psi = p \cos \phi, \quad p\rho = a^2 \sin^2 \phi + b^2 \cos^2 \phi,$$

whence
$$ax = (a^2 - b^2) \cos^3 \phi,$$

$$by = (a^2 - b^2) \sin^3 \phi.$$

Accordingly the equation of the evolute is

$$(ax)^{\frac{2}{3}} + (by)^{\frac{2}{3}} = (a^2 - b^2)^{\frac{2}{3}}.$$

The form of this equation shows that we cannot deduce properties of the four-cusped hypocycloid by supposing an ellipse to degrade into a circle; but if the equation be written in the form

$$(x/A)^{\frac{2}{3}} + (y/B)^{\frac{2}{3}} = 1,$$

properties of the evolute may be deduced from those of the hypocycloid by orthogonal projection.

337. In the figure, let CZ be the perpendicular from the centre of an ellipse on to the normal at P, O the centre of curva-

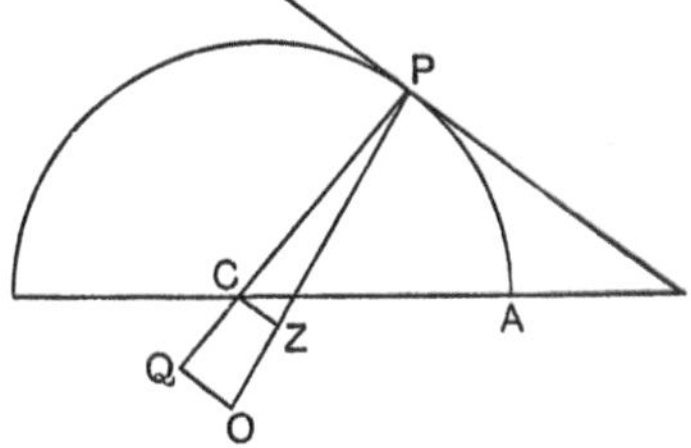

ture, and OQ perpendicular to PO. Then if $CZ = p'$, $ZCA = \frac{1}{2}\pi - \psi$, it follows from the properties of the ellipse that

$$p' = CZ = \frac{(a^2 - b^2) \sin \psi \cos \psi}{(a^2 \cos^2 \psi + b^2 \sin^2 \psi)^{\frac{1}{2}}},$$

and therefore the pedal of the evolute is the sextic curve

$$(a^2 y^2 + b^2 x^2)(x^2 + y^2)^2 = (a^2 - b^2)^2 \, x^2 y^2,$$

whilst the orthoptic locus is the sextic curve given at the end of § 68.

338. *To prove that the radius of curvature is equal to $3OQ$.*

We shall first prove that if ρ be the radius of curvature of any curve, and ψ the angle which the normal makes with the axis of x, then the radius of curvature of the evolute is $-d\rho/d\psi$.

Let the accented letters refer to the evolute; then

$$ds' = d\rho, \quad \psi' = \tfrac{1}{2}\pi - \psi,$$

therefore
$$\rho' = ds'/d\psi' = -d\rho/d\psi.$$

Again, since
$$p' = CZ = dp/d\psi,$$

we obtain
$$\rho' = -p'\, d\rho/dp.$$

From the properties of the ellipse
$$p^3\rho = a^2 b^2,$$

whence
$$-d\rho/dp = 3\rho/p,$$

also
$$p' = p \cdot OQ/\rho.$$

Accordingly
$$\rho' = 3OQ.$$

339. *If the tangent at any point P intersects the curve in Q and R, the locus of the point of intersection of the tangents at Q and R is an ellipse.*

If the equation of the curve is

$$(x/a)^{\frac{2}{3}} + (y/b)^{\frac{2}{3}} = 1 \quad \ldots\ldots\ldots\ldots\ldots\ldots(1),$$

the equation
$$\frac{h}{a^{\frac{2}{3}} x^{\frac{1}{3}}} + \frac{k}{b^{\frac{2}{3}} y^{\frac{1}{3}}} = 1 \quad \ldots\ldots\ldots\ldots\ldots\ldots(2),$$

is that of a curve which passes through the points of contact of the tangents from (h, k); also the equation of the tangent at any point (f, g) is

$$\frac{x}{a^{\frac{2}{3}} f^{\frac{1}{3}}} + \frac{y}{b^{\frac{2}{3}} g^{\frac{1}{3}}} = 1 \quad \ldots\ldots\ldots\ldots\ldots\ldots(3).$$

Let (h, k) be the point of intersection of the tangents at Q and R; then since (x, y) and (f, g) are points on the evolute, we may write

$$x = a \cos^3 \phi, \quad y = b \sin^3 \phi,$$
$$f = a \cos^3 \psi, \quad g = b \sin^3 \psi,$$

and (2) and (3) become

$$\frac{h}{a}\sec\phi + \frac{k}{b}\operatorname{cosec}\phi = 1 \quad\dots\dots\dots\dots(4),$$

$$\frac{\cos^3\phi}{\cos\psi} + \frac{\sin^3\phi}{\sin\psi} = 1 \quad\dots\dots\dots\dots\dots(5).$$

Let $\lambda = \tan\phi$, then (4) may be written in the form

$$\frac{h^2\lambda^4}{a^2} + \frac{2hk}{ab}(\lambda^3+\lambda) + \left(\frac{h^2}{a^2}+\frac{k^2}{b^2}-1\right)\lambda^2 + \frac{k^2}{b^2} = 0 \quad\dots\dots(6).$$

This equation determines the value of ϕ at the four points of contact of the tangents from $(h,\ k)$. Equation (5) determines the value of λ in terms of ψ, and is a sextic equation in λ. The sextic obviously contains $(\lambda-\tan\psi)^2$ as a factor; and the quartic factor will be found to be

$$\lambda^4 + 2(\lambda^3+\lambda)\tan\psi + \tan^2\psi = 0 \quad\dots\dots\dots\dots(7).$$

Since (6) and (7) are satisfied by the same value of λ, it follows that

$$\frac{h^2}{a^2} + \frac{k^2}{b^2} = 1,$$

$$\tan\psi = ak/bh.$$

The first equation determines the locus of $(h,\ k)$ which is an ellipse; whilst the second gives the value of ψ in terms of $(h,\ k)$. When the evolute degrades into a four-cusped hypocycloid, the locus becomes a circle.

The reciprocal theorem is as follows :—

If two tangents be drawn to the reciprocal curve from a point on itself, the envelope of the chord of contact is an ellipse, which becomes a circle when the curve is the reciprocal polar of a four-cusped hypocycloid.

340. *To find the tangential equation of the evolute of the evolute of an ellipse.*

The equation of the normal to the evolute at O is

$$x\cos\psi + y\sin\psi = p - \rho,$$

whence $\qquad\qquad \xi = \frac{\cos\psi}{p-\rho}, \qquad \eta = \frac{\sin\psi}{p-\rho};$

accordingly
$$a^2\xi^2 + b^2\eta^2 = \frac{p^2}{(\rho - p)^2},$$

$$\xi^2 + \eta^2 = \frac{1}{(\rho - p)^2}.$$

Therefore
$$a^2\xi^2 + b^2\eta^2 = p^2(\xi^2 + \eta^2),$$

also
$$(\rho - p)^2 = \frac{1}{p^2}\left(\frac{a^2 b^2}{p^2} - p^2\right)^2,$$

therefore
$$p^2(\rho - p)^2 = \frac{(a^2 - b^2)^2 (b^2\eta^4 - a^2\xi^4)^2}{(a^2\xi^2 + b^2\eta^2)^2(\xi^2 + \eta^2)^2},$$

whence the required equation is

$$(a^2\xi^2 + b^2\eta^2)^3 = (a^2 - b^2)^2 (b^2\eta^4 - a^2\xi^4)^2,$$

and is therefore a curve of the eighth class.

341. The evolute of an ellipse can also be generated in the following manner, which can be proved directly or by orthogonally projecting a four-cusped hypocycloid on a plane parallel to one of the cuspidal tangents.

From any point P on an ellipse draw perpendiculars PM, PN to the major and minor axes; draw ME parallel to the tangent at P to meet CP in E; draw EL perpendicular to the major axis cutting MN in R. Then MN is the tangent at R to the evolute of an ellipse.

342. *The evolute of an ellipse has two real single foci, which are the foci of the ellipse.*

We have shown in § 65 that the tangential equation of the curve

$$(x/A)^{\frac{2}{3}} + (y/B)^{\frac{2}{3}} = 1$$

is
$$\frac{1}{A^2\xi^2} + \frac{1}{B^2\eta^2} = 1,$$

whence if (α, β) be the coordinates of any focus

$$(\alpha + \iota\beta)^2\left(\frac{1}{A^2} - \frac{1}{B^2}\right) = 1 \dots\dots\dots\dots\dots(1).$$

But
$$A = (a^2 - b^2)/a, \quad B = (a^2 - b^2)/b,$$

whence (1) becomes
$$(\alpha + \iota\beta)^2 = a^2 - b^2,$$

and therefore
$$\alpha = \pm(a^2 - b^2)^{\frac{1}{2}}, \quad \beta = 0.$$

Equation (1) may be written

$$\alpha + \iota\beta = \pm \frac{AB}{(B^2 - A^2)^{\frac{1}{2}}},$$

which shows that the two real single foci of a four-cusped hypocycloid are at infinity in opposite directions on the axis of x.

The Involute of a Circle.

343. Let PQ be the tangent at any point Q of a circle of radius a; and let $QP = QA$, where A is a fixed point; then the

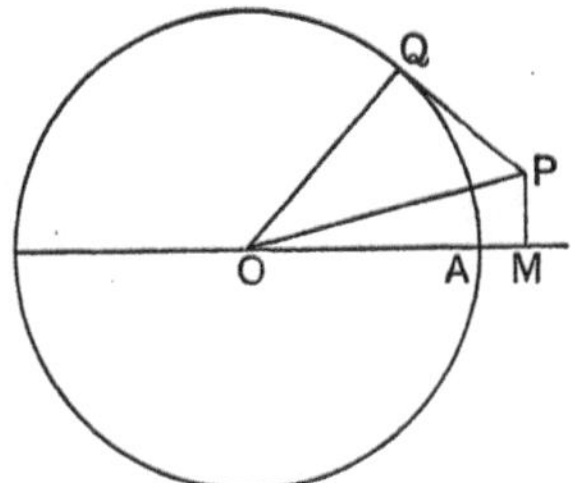

locus of P is the involute of a circle.

Let $$OP = r, \quad POM = \theta, \quad QOA = \phi.$$

Then $$r\cos(\phi - \theta) = a, \quad a\phi = PQ = r\sin(\phi - \theta),$$

whence, eliminating ϕ, we get

$$(r^2 - a^2)^{\frac{1}{2}} = a\theta + \cos^{-1} a/r.$$

The coordinates of P are also given by the equations

$$x = a\cos\phi + a\phi\sin\phi,$$

$$y = a\sin\phi - a\phi\cos\phi.$$

Since PQ is equal to the perpendicular from the centre O on to the tangent at P to the involute, the p and r equation is

$$r^2 = a^2 + p^2,$$

and the tangential polar equation is

$$p = a\phi,$$

whence, as will appear in § 349, the pedal of the involute is the spiral of Archimedes.

15—2

344. *If the involute of a circle roll on a straight line, the locus of the centre of the circle of which it is the involute is a parabola.*

If (x, y) be the coordinates of the locus of O referred to the point on the straight line with which A was initially in contact, we have by § 314,

$$y = p,$$

$$\tan \phi = \frac{dx}{dy} = \frac{p}{(r^2 - p^2)^{\frac{1}{2}}} = \frac{y}{a},$$

whence
$$y^2 = 2ax.$$

The Catenary.

345. The catenary is the curve in which a flexible inelastic string hangs, when suspended from two points under the action of gravity.

To find the equation of the catenary.

Let A and B be the points of suspension, C the lowest point of the string. Let T be the tension at any point P of the curve; wc the tension at the lowest point C, where w is the weight of a

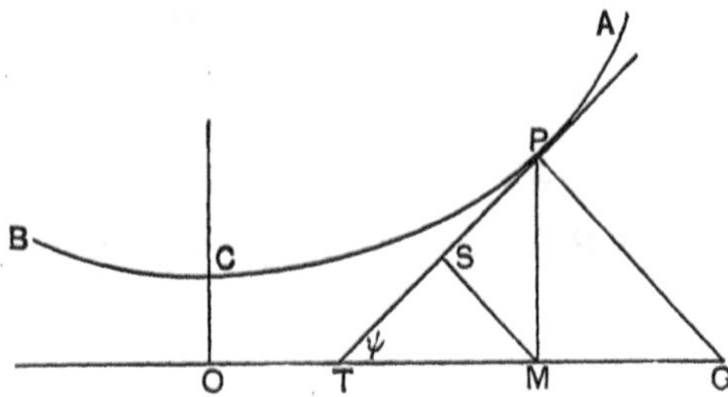

unit of length; s the length of the arc CP; ψ the angle which the tangent at P makes with the horizontal line OG.

The equations of equilibrium of CP are

$$T \cos \psi = wc, \quad T \sin \psi = ws,$$

whence
$$s = c \tan \psi \dots\dots\dots\dots\dots\dots\dots\dots\dots (1),$$

which is the intrinsic equation of the curve.

Now
$$\sec \psi = \frac{ds}{dx} = c \sec^2 \psi \frac{d\psi}{dx},$$

whence
$$x = c \log (\tan \psi + \sec \psi),$$

from which we obtain

$$\frac{dy}{dx} = \tan \psi = \sinh x/c,$$

and therefore
$$y = c \cosh x/c \dots\dots\dots\dots\dots\dots\dots\dots\dots(2),$$

which is the Cartesian equation.

In the figure let OG be a horizontal line such that $OC = c$; then OG is called the directrix of the catenary. Let the tangent, normal and ordinate at P meet OG in T, G and M. Draw MS perpendicular to PT. Then

$$\tan \psi = \frac{dy}{dx} = \frac{1}{c} \frac{dy}{d\psi} \cos \psi,$$

whence $\qquad\qquad y = c \sec \psi.$

But $\qquad\qquad y = PM = MS \sec \psi,$

therefore $\qquad\qquad MS = c.$

Also $\qquad\qquad s = c \tan \psi = MS \tan \psi = PS,$

which shows that the locus of S is the involute of the catenary.

Again if ρ be the radius of curvature at P,

$$\rho = \frac{ds}{d\psi} = c \sec^2 \psi$$

$$= PM \sec \psi = PG,$$

whence the centre of curvature of P is at a point O' on GP produced such that $PO' = PG$.

If a catenary revolve about its directrix, the line PG is the radius of curvature of the circular section of the surface of revolution thereby generated, whilst PO' is the radius of curvature of the meridian section; and it is known from solid geometry that these are the two principal sections of the surface. Hence the surface generated by the revolution of a catenary about its directrix belongs to the class of surfaces which have their two principal radii of curvature equal and opposite.

346. It can also be shown by the method of § 314 that if *a parabola roll on a straight line, the locus of its focus is a catenary.* If however the conic is a central one, the locus of the focus is a more complicated curve*.

* Besant, *Notes on Roulettes*, p. 47.

347. The involute of a catenary is a curve called the Tractrix or Tractory. To find its equation, we observe that the curve is the locus of S, whence if (x, y) denote the coordinates of S,

$$y = c \cos \psi$$

and

$$SP = c \tan \psi = \frac{ds}{d\psi},$$

whence

$$s = c \log \sec \psi,$$

which is the intrinsic equation of the curve.

Again

$$\frac{dx}{ds} = \sin \psi,$$

whence

$$\frac{dx}{d\psi} = c (\sec \psi - \cos \psi),$$

therefore

$$x = c \log (\tan \psi + \sec \psi) - c \sin \psi.$$

Eliminating ψ we get

$$x + (c^2 - y^2)^{\frac{1}{2}} = c \log \{c/y + (c^2/y^2 - 1)^{\frac{1}{2}}\}.$$

348. There are a variety of other curves of an analogous kind, such as the catenary of equal strength and the catenary formed by an elastic string; but for the discussion of these curves we must refer to works on Statics. There is also a curve called the *elastica*, which is the curve assumed by an elastic wire whose natural form is a straight line, which is bent in its own plane without torsion. This curve is discussed in my *Elementary Treatise on Hydrodynamics and Sound*, and by putting $\alpha = 0$ in (7) of § 139 of the second edition of that treatise, it can be shown that if the string joining the ends of an elastic wire is of such a length that it is the normal at the two extremities of the wire, the equation of the elastica is of the form

$$\frac{dx}{dy} = \frac{y^2}{(4a^4 - y^4)^{\frac{1}{2}}},$$

the string being the axis of x, and the origin its middle point. This result, combined with § 314, leads to the theorem that :—
If a rectangular hyperbola roll on a straight line, the locus of its centre is an elastica.

The equation of the curve assumed by a rectangular piece of flexible and inextensible material filled with liquid, whose sides AB, CD are fastened to the sides of a box, and whose other sides fit the box so closely that the liquid cannot escape, was first investigated by James Bernoulli*, who called it the lintearia. The curve is, however, the same as an elastica.

Spirals.

349. We shall conclude this chapter with an account of certain spiral curves.

The *equiangular spiral*, or the logarithmic spiral as it is sometimes called, is a curve such that the tangent at any point makes a constant angle with the radius vector.

To find its equation, let α be the angle of the spiral, then

$$r\frac{d\theta}{dr} = \tan\alpha,$$

whence $\qquad\qquad r = a\epsilon^{\theta\tan\alpha}.$

The p and r equation of the curve is

$$p = r\sin\alpha,$$

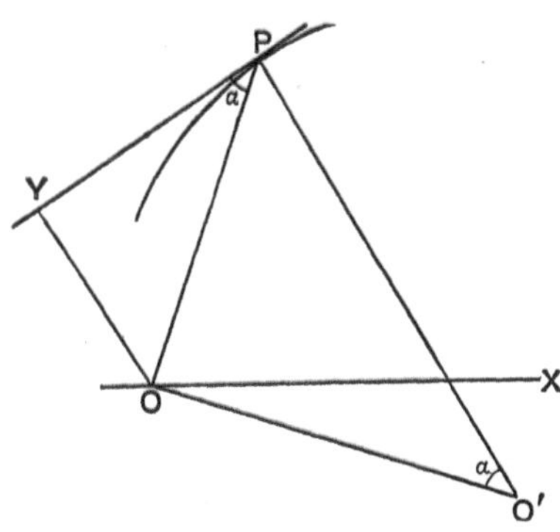

whence $\qquad\qquad \rho = r\,\mathrm{cosec}\,\alpha.$

Hence if from the origin O a straight line be drawn perpendicular to OP to meet the normal in O', then O' is the centre of curvature of P.

Let $OO' = r'$, $O'OX = \theta'$; then

$$r' = r\cot\alpha, \quad \theta' = \theta - \tfrac{1}{2}\pi,$$

* Besant's *Hydromechanics*, Ch. VIII.; Walton's *Hydrostatical Problems*, p. 207.

whence $\qquad r' = a \cot \alpha \cdot \epsilon^{(\frac{1}{2}\pi + \theta')\tan \alpha},$

and therefore the evolute is a similar spiral.   Also the pedal is

$$OY = a \sin \alpha \cdot \epsilon^{\theta \tan \alpha},$$

which is a similar spiral.

350.   The *spiral of Archimedes* is the curve $r = a\theta$; and by § 343 it is the pedal of the involute of a circle.

The *hyperbolic spiral* is the curve $r\theta = a$, and is the reciprocal polar of the involute of a circle.   It has an asymptote whose distance from the initial line is a.

The *lituus* is the curve $r^2\theta = a^2$, and the initial line is an asymptote.   These last three spirals are included in the equation $r = a\theta^n$.

CHAPTER XII.

THEORY OF PROJECTION.

351. THE theory of projection is explained in treatises on Conics, but since it affords a powerful method of deducing general properties of curves from those of curves of a more simple form we shall explain its leading features, and then apply it to deduce properties of cubic and quartic curves.

If a curve S be drawn in any plane z, and if with any point V as vertex a cone be described whose generators pass through S, the curve of intersection of the cone with any plane z' is called the projection of S.

If any straight line through the vertex cut the planes z and z' in P and P', these points are called *corresponding points*. In other words, the projection of P on the plane z' is called the point corresponding to P.

The projection of any straight line is obviously another straight line. Also if any straight line cuts a curve in n points $P_1, P_2, \ldots P_n$, its projection will cut the projection of the curve in n corresponding points $P_1', P_2', \ldots P_n'$. But since every straight line cuts a curve of the nth degree in n real or imaginary points, it follows that the projections of the line and curve cut one another in the same number of points. Hence the projection of a curve of the nth degree is another curve of the same degree; also a tangent to a curve projects into a tangent to the projection of the curve, and every singularity on a curve projects into the same singularity on the projected curve.

It can be shown by elementary geometry that the projection of a triangle on any parallel plane is a similar triangle; from which it follows that the projection of a polygon on a parallel plane is a similar polygon. Also since every curve may be regarded as the limit of a polygon, it follows that the projection of any curve on a parallel plane is a similar curve.

Through the vertex V draw a plane parallel to z cutting the plane z' in a straight line l'; then the line l' is the projection on z' of the line at infinity on z. Similarly if a plane through V parallel to z' cut z in l, the projection of l is the line at infinity on z'. Hence any line on z can be projected to infinity, whilst the line at infinity on z can be projected into any line on z'. From this result it follows that the properties of curves having singularities at a finite distance from the origin can be deduced from those of curves having the corresponding singularities at infinity; whence the properties of quartics having a pair of imaginary nodes or cusps may be deduced from the known properties of bicircular quartics and cartesians. We shall also show that the properties of a curve having a pair of real nodes or cusps may be deduced from those of a curve having a pair of imaginary ones; from which it follows that the properties of curves having imaginary singularities can be deduced from those having real ones and *vice versâ*. The theory of projection can also be employed to examine whether a curve has a singularity at infinity, since any point on the line at infinity can be projected into the origin.

All straight lines which are parallel to the line of intersection of the plane of projection with the original plane, project into parallel straight lines; but parallel lines which are not parallel to this line project into lines passing through a point, which is the intersection of the projection of the line at infinity with the common line of intersection of the planes passing through the parallel straight lines and the vertex.

352. *A projection introduces five independent constants.*

Since the projection of a curve on a parallel plane is a similar curve, we may without loss of generality suppose the plane of projection to pass through a point O in the original plane, which we shall choose as the origin of three rectangular axes Ox, Oy, Oz, of which Oz is perpendicular to the original plane. When the equation of a curve referred to Ox, Oy is given, its projection is completely determined by the coordinates (ξ, η, ζ) of V the vertex, the inclination ϵ of the plane of projection to the original plane, and the angle η which their line of intersection makes with the axis of x or y, which are the five constants in question.

353.   *Any given triangle can be projected into any other given triangle, in such a manner that any given point P in the plane of the first triangle corresponds to any given point Q in the plane of the second.*

Let ABC be the given triangle; let A be the origin, AB the axis of x, and let two lines through A in and perpendicular to the plane ABC be the axes of y and z.

Let ABC' be a triangle similar to the second triangle, such that the base AB is equal to that of the original triangle; and place this triangle so that the bases AB coincide, whilst the plane ABC' makes an angle ϵ with the plane z.

Let $(f, g, 0)$ be the coordinates of C, and $(p, q, 0)$ those of any point P in the plane z. Then if accented letters denote the coordinates of the corresponding points in the plane ABC', the coordinates of C' and P' are f', $g' \cos \epsilon$, $g' \sin \epsilon$, and p', $q' \cos \epsilon$, $q' \sin \epsilon$ respectively. Hence the equations of CC' and PP' are

$$\frac{x - f}{f' - f} = \frac{y - g}{g' \cos \epsilon - g} = \frac{z}{g' \sin \epsilon}$$

and

$$\frac{x - p}{p' - p} = \frac{y - q}{q' \cos \epsilon - q} = \frac{z}{q' \sin \epsilon}.$$

The conditions that the projection should be possible require that the lines CC' and PP' should intersect at a point V, which is the vertex. Now the foregoing equations are sufficient to determine (x, y, z), which gives the point V, and also the angle ϵ which the plane of projection makes with the plane of the triangle. If therefore a line $A'B'$ be drawn in the plane VAB parallel to AB and equal to the side $A'B'$ of the second triangle, the section of the pyramid $VABC$ by a plane through $A'B'$ parallel to the plane ABC' will be a triangle equal to the second given triangle; also the point where VP cuts the plane of the latter triangle will be the given point Q.

The preceding theorem shows that any quadrilateral can be projected into a square. For let $ABCD$ be the quadrilateral, and $A'B'C'D'$ a square; then the triangle ABC can be projected into the isosceles right-angled triangle $A'B'C'$ so that the point D' corresponds to D.

354. We shall now give the analytical formulae for projection, and shall first suppose that the axis of y is the line of intersection of the two planes, and that the axis of x in the plane of projection is the projection of the original axis of x.

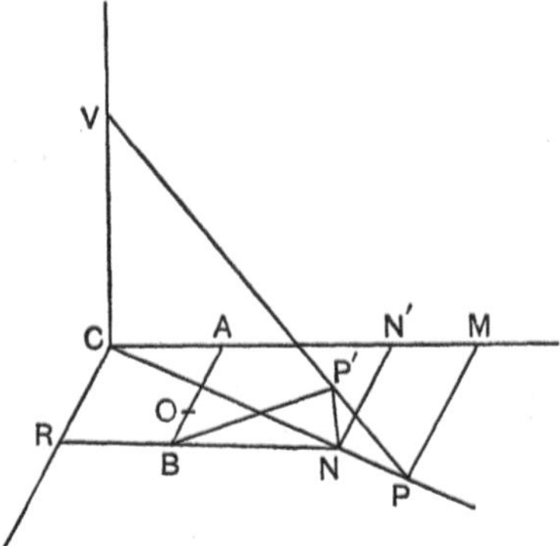

Let AB be the line of intersection of the plane of projection $AP'B$ with the original plane CPM; let V be the vertex, VC perpendicular to the plane CPM; CM, CR parallel to the axes of x and y. Let (x, y) be the coordinates of P referred to the origin O, and let OB be the axis of y; also let (x', y') be the coordinates of the corresponding point P' referred to axes in the plane $OP'B$, of which OB is the axis of y', and a line in this plane perpendicular to OB is the axis of x'. Let (ξ, η, ζ) be the coordinates of V; $\epsilon = $ angle $P'BN$. Then

$$\frac{P'N}{VC} = \frac{PN}{PC} = \frac{MN'}{MC} = \frac{PM - NN'}{PM};$$

also since $OA = -\eta$, $AC = -\xi$, $VC = \zeta$,

$$\frac{x' \sin \epsilon}{\zeta} = \frac{x - x' \cos \epsilon}{x - \xi} = \frac{y - y'}{y - \eta};$$

accordingly

$$\left. \begin{aligned} x &= \frac{x' (\zeta \cos \epsilon - \xi \sin \epsilon)}{\zeta - x' \sin \epsilon} \\[2mm] y &= \frac{y'\zeta - x'\eta \sin \epsilon}{\zeta - x' \sin \epsilon} \end{aligned} \right\} \quad \dots\dots\dots\dots\dots (1),$$

which are the required formulae. More complicated ones can be obtained by the usual methods for the transformation of coordinates.

The alternative formulae are

$$x' = \frac{x\zeta}{\zeta \cos \epsilon - \xi \sin \epsilon + x \sin \epsilon}$$
$$y' = \frac{x\eta \sin \epsilon + y\,(\zeta \cos \epsilon - \xi \sin \epsilon)}{\zeta \cos \epsilon - \xi \sin \epsilon + x \sin \epsilon} \qquad \Bigg\} \ \ldots\ldots\ldots\ldots(2).$$

355. *Any two points can be projected into the circular points at infinity and vice versâ.*

Let P and Q be two points in the plane z'. Take a point O as origin in the line of intersection of the planes z and z', such that OPQ is an isosceles triangle whose vertex is O, and let the equations of OP, OQ be $y' = \pm mx'$; and let the equation of PQ be $x' = l$.

Through PQ draw a plane parallel to the plane z, and let the vertex V be a point in this plane whose coordinates referred to O are ξ, 0, $l \sin \epsilon$. Then the equation of the projection of OP on the plane z is

$$y\,(l \cos \epsilon - \xi) = mlx,$$

and the projection of P is the intersection of this line with the line at infinity on z. If this point is the circular point $x = \iota y$, we must have

$$l \cos \epsilon - \xi = \iota ml.$$

Let $\epsilon = \tfrac{1}{2}\pi - \iota\beta$; then $\cos \epsilon = \iota \sinh \beta$, whence

$$\xi = \iota l\,(\sinh \beta - m),$$
$$\eta = 0,$$
$$\zeta = l \cosh \beta.$$

Accordingly (2) become

$$x' = \frac{lx}{\iota ml + x}$$
$$y' = \frac{\iota mly}{\iota ml + x} \qquad \Bigg\} \ \ldots\ldots\ldots\ldots\ldots\ldots(3),$$

which give

$$x = \frac{\iota mlx'}{l - x'}$$
$$y = \frac{ly'}{l - x'} \qquad \Bigg\} \ \ldots\ldots\ldots\ldots\ldots\ldots(4).$$

Equations (3) furnish formulae by means of which *any* two real points can be projected into the circular points; whilst (4) furnish formulae for projecting the circular points into a pair of real points. It will be observed that the projection is imaginary.

356. It appears from the foregoing articles that all properties of a curve which do not involve the magnitude of lines or angles are capable of being generalized by projection; whilst those which depend upon the magnitude of such quantities cannot be generalized by this method, except in certain special cases. Hence the properties of curves are frequently divided into two classes called *metric* and *descriptive*, according as they do or do not involve the magnitudes of lines and angles. But properties relating to lines cutting one another at a constant angle, which include theorems concerning orthoptic loci, can be generalized by projection as we shall proceed to show.

357. It is shown in treatises on Conics that the anharmonic ratio of a pencil remains altered by projection; from which it follows that if a line is divided harmonically, its projection is also divided harmonically. It is also known that the four straight lines α, β, $\alpha + k\beta$, $\alpha - k\beta$, which pass through the point C of the triangle of reference, form a harmonic pencil; from which it follows that if through a point O any two lines be drawn at right angles to one another, these lines, together with the pair of imaginary lines drawn from O to the circular points at infinity, form a harmonic pencil. Hence properties of lines intersecting at right angles can be projected into harmonic properties.

Again, if two straight lines intersect at a constant angle, the anharmonic ratio of the pencil formed by them and the two lines drawn from their point of intersection to the circular points is also constant. Hence properties connected with lines which intersect at a constant angle can be generalized by projection.

358. The theory of projection also shows that the partivity of a curve may be equal to, but cannot exceed, its degree. To fix our ideas, consider an elliptic limaçon with two real points of inflexion. If any line in the plane of the curve be projected to infinity, the projected curve will be quadripartite, tripartite, bipartite or unipartite according as the line intersects the curve in (i) four real points, (ii) two distinct and two coincident real points, (iii) two real and two imaginary points, and (iv) four

imaginary points respectively. Also by means of an imaginary projection an elliptic limaçon with two imaginary points of inflexion can be projected into a quadripartite quartic.

359. In order to apply the theory of projection with advantage, the first step is to draw up a table of the simplest form or forms which curves of any given species can assume. The next step is to investigate the properties of these simple forms by any convenient method, and then to generalize those which are capable of projection. We shall therefore proceed to examine some of the simplest forms of cubic and quartic curves, and shall incidentally show that a variety of results, some of which are known whilst others are probably new, may be deduced from the properties of various well known curves.

Cubic Curves.

360. *Any nodal cubic can be projected into the logocyclic curve; and every cuspidal cubic into a cissoid.*

The equation of the logocyclic curve in its simplest form is $x(x^2 + y^2) = a(x^2 - y^2)$, and therefore contains one constant. Transfer the origin to any point in the plane (x, y) and two more constants will be introduced, which make three. Project on any plane passing through the new origin, and five more constants will be introduced making eight, which is the number of independent constants which the general equation of a nodal cubic contains.

The logocyclic curve has one real point of inflexion I at infinity, and the asymptote is the inflexional tangent ; hence the tangent at the vertex A is the tangent drawn from the real point of inflexion to the curve. Now since the nodal tangents bisect the angles between OI and OA (which are at right angles) it follows that these lines together with the nodal tangents form a harmonic pencil; hence the possibility of projecting any nodal cubic into the logocyclic curve, depends upon the following theorem, which we shall proceed to prove.

361. *If from any point of inflexion A of a cubic whose node is B, a tangent be drawn touching the curve at C, the lines BA, BC together with the nodal tangents form a harmonic pencil.*

If ABC be the triangle of reference, the equation of the cubic is

$$\alpha^2 (\mu\beta + \nu\gamma) + m\beta\gamma^2 = 0,$$

and the nodal tangents at B are $\mu\alpha^2 + m\gamma^2 = 0$, which together with BA, BC form a harmonic pencil.

It might however have happened, in the case of a nodal cubic, that the four lines in question did not form a harmonic pencil, in which case it would not be possible to project every nodal cubic into the logocyclic curve. We shall have examples of this in the case of quartic curves; and it is necessary to warn the reader that counting the constants is not always a safe process, since the condition thereby furnished, although a necessary one, is not always a sufficient one.

It can be shown in a similar manner that every cuspidal cubic can be projected into a cissoid; also since the reciprocal curve is a cubic of the third class, properties of one cuspidal cubic can be deduced from those of another by reciprocation.

362. We shall now give some examples of the projective properties of nodal cubics. From the theorems of §§ 131 and 127 we obtain :—

(i) *From the point of contact A of the tangent from any point of inflexion I of a nodal cubic, draw a chord APP'. Join IP' cutting the cubic in p. Then the tangents at P and p intersect on the curve.*

(ii) *If from a point of inflexion of a nodal cubic a tangent be drawn, and through the point of contact any chord be drawn, the locus of the point of intersection of the tangents at the other two points where the chord cuts the curve is a cuspidal cubic, whose cusp coincides with the node of the cubic.*

The reciprocal theorem is the following :—

(iii) *Let any cuspidal tangent of a tricuspidal quartic cut the curve at O; from any point on the tangent at O draw a pair of tangents to the quartic. Then the envelope of their chord of contact is a cuspidal cubic.*

363. *Every anautotomic cubic can be projected into the circular cubic*

$$x (x^2 + y^2 \pm a^2) = b (x^2 - y^2).$$

For if the origin be transferred to any point two new constants are introduced, and projection adds five more making altogether nine, which is the number of independent constants which the general equation of a cubic curve contains.

364. There is another form of an anautotomic cubic which is occasionally useful, in which the circular points are points of inflexion. This form, so to speak, localizes the circular points, and thereby enables properties connected with points of inflexion to be deduced from those of the circular points. By § 123 the equation in question may be written

$$(x^2 + y^2)(x + a) + c^3 = 0,$$

where $x + a = 0$ is the tangent at the real point of inflexion at infinity. By transferring the origin to one of the points where the axis of x cuts the curve, the equation may be written

$$x(x^2 + y^2) + (3\beta - 2\alpha)\,x^2 + \beta y^2 + (\beta - \alpha)(3\beta - \alpha)\,x = 0,$$

where by § 49, $x + \beta = 0$ is the tangent at the real point of inflexion at infinity. If the origin is a node, $\alpha = 3\beta$ and the equation of the curve becomes

$$x(x^2 + y^2) - 3\beta x^2 + \beta y^2 = 0,$$

which is the trisectrix of Maclaurin.

365. It may be shown by the method of § 70, that an ellipse intersects the lines $y = \pm\, \iota b x/a$ at two imaginary points at infinity, which may be called the *elliptic points at infinity*; hence a theory of elliptic cubics exists of the same character as that of circular cubics. The equation of such a cubic is

$$u_1\,(x^2/a^2 + y^2/b^2) + u_2 + u_1 + u_0 = 0,$$

and the properties of these curves may be derived either directly or by projecting those of circular cubics. When the two points coincide, we obtain a *parabolic point at infinity*, which is the point of contact of a parabola with the line at infinity; whilst the *hyperbolic points* at infinity are the two real points where a hyperbola touches its asymptotes.

Quartic Curves.

366.   We shall first consider the projection of a quartic with three double points.

We have already shown that any two imaginary points may be projected into a pair of real points; if therefore a quartic has a pair of imaginary nodes or cusps, the curve may be projected into a quartic having a pair of real nodes or cusps; also the triangle whose vertices are the double points may be projected into an equilateral triangle, such that any given point in its plane coincides with any given point in the plane of the original triangle.   Whence :—

Any tricuspidal quartic may be projected into a three-cusped hypocycloid or into a cardioid.

The projection may be accomplished in the first case by projecting the cuspidal triangle into a *real* equilateral triangle, whose centre of gravity coincides with the point of intersection of the three cuspidal tangents; whilst in the second case, two of the cusps must be projected into the circular points at infinity.

367.   We shall now give some examples.

We have shown in § 325 that:—*If any tangent to a three-cusped hypocycloid cuts the curve in P and Q, the tangents at these points intersect at right angles on a circle which touches the hypo-cycloid at three points; also the line at infinity is the only double tangent, and the points of contact are the circular points.*   Whence by projection:—

(i) *If any tangent to a tricuspidal quartic cuts the curve in P and Q, the tangents at these points intersect on a conic, which (α) touches the quartic at three points; (β) intersects it at the points of contact of the double tangent; (γ) also the point of intersection of the three cuspidal tangents is the pole of the double tangent with respect to the conic.*

From § 357, it follows that:—

(ii) *The two tangents, together with the lines joining their point of intersection with the points of contact of the double tangent, form a harmonic pencil.*

In § 301 we proved that :—

Tangents at the extremities of a chord drawn through the real cusp of a cardioid intersect at right angles on a circle, whose centre is the triple focus, and which touches the cardioid at the point where the cuspidal tangent intersects it.

Whence by projection :—

(iii) *If through any cusp of a tricuspidal quartic a chord be drawn cutting the curve in P and Q, the tangents at these points intersect on a conic which (α) touches the quartic at the point of intersection of the cuspidal tangent with the quartic; (β) passes through the other two cusps of the quartic; (γ) also the point of intersection of the cuspidal tangents to the quartic is the pole of the line joining the other two cusps with respect to the conic.*

(iv) *The tangents at P and Q, together with the lines joining their point of intersection with the other two cusps, form a harmonic pencil.*

From the theorem that the evolute of a cardioid is another cardioid, we obtain :—

(v) *Through any point P on a tricuspidal quartic draw two straight lines to a pair of cusps; draw the tangent at P and the harmonic conjugate of these three straight lines. Then the envelope of the harmonic conjugate is another tricuspidal quartic, two of whose cusps coincide with the pair of cusps of the original quartic.*

368. *Any quartic having a node and a pair of cusps can be projected into a limaçon.*

For the quartic can be projected into a curve having a node at the origin and a pair of cusps at the circular points, and the limaçon is the only quartic having these singularities at the above mentioned points.

The ninth species of quartics, of which the limaçon is one of the simplest forms, is of great interest owing to the fact that such curves reciprocate into quartics of the fourth class. We may therefore deduce properties of quartics belonging to this species by first projecting those of the limaçon, and then reciprocating the projected curve with respect to any origin.

369. In the limaçon, the nodal tangents, the line joining the node with the triple focus and the line drawn through the node

parallel to the double tangent form a harmonic pencil. The two last lines being parallel, intersect on the line at infinity; whence the projective theorem is:—

(i) *In any quartic of the ninth species, the nodal tangents, together with the lines joining the node to the point of intersection of the cuspidal tangents and to that of the double tangent with the line drawn through the cusps, form a harmonic pencil.*

A direct proof of this theorem may be given as follows. Let A be the node, B and C the cusps, D the point of intersection of the cuspidal tangents and E the point where the double tangent intersects BC. The equation of the quartic is

$$\lambda^2\beta^2\gamma^2 + \mu^2\gamma^2\alpha^2 + \nu^2\alpha^2\beta^2 - 2\alpha\beta\gamma\,(l\alpha + \lambda\nu\beta + \lambda\mu\gamma) = 0 \ldots\ldots(1),$$

whence the equation of AD is

$$\nu\beta - \mu\gamma = 0 \ \ldots\ldots\ldots\ldots\ldots\ldots(2).$$

By § 191, the equation of the double tangent is

$$(l + \mu\nu)\,\alpha + 2\lambda\,(\nu\beta + \mu\gamma) = 0,$$

whence the equation of AE is

$$\nu\beta + \mu\gamma = 0 \ \ldots\ldots\ldots\ldots\ldots\ldots(3).$$

The equation of the nodal tangents is

$$\nu^2\beta^2 + \mu^2\gamma^2 - 2l\beta\gamma = 0,$$

whence writing $l = \mu\nu \cosh\theta$, the equation of the nodal tangents becomes

$$\left.\begin{array}{l} \nu\beta - \mu\gamma\epsilon^{\theta} \ = 0 \\ \nu\beta - \mu\gamma\epsilon^{-\theta} = 0 \end{array}\right\}\ldots\ldots\ldots\ldots\ldots(4),$$

which shows that the four lines (2), (3) and (4) form a harmonic pencil.

370. From the theorem of § 287 we obtain by projection:—

(ii) *Tangents at the extremities of any chord through the node of the quartic intersect on a nodal cubic, which passes through the cusps of the quartic.*

Reciprocating this theorem, we obtain:—

(iii) *From any point on the double tangent to the quartic draw a pair of tangents to the curve; then the envelope of the chord of contact is a tricuspidal quartic, which touches the two inflexional tangents of the first quartic.*

The external focus F_1 is the only single focus of an elliptic limaçon, and is therefore the intersection of the tangents drawn from the circular points to the curve; whence by projecting the theorem of § 286 we obtain :—

(iv) *From each cusp draw a tangent to the quartic, and through their point of intersection draw a chord cutting the quartic in two real points ; then the locus of the point of intersection of the tangents at these points is a cuspidal cubic which passes through the cusps of the quartic, and whose cusp coincides with the node of the quartic.*

Reciprocating, we obtain :—

(v) *Let O be any point on the line joining the two points where the two inflexional tangents intersect the curve ; from O draw a pair of tangents to the quartic which touch the curve at two real points ; then the envelope of the chord of contact is a cuspidal cubic which touches the inflexional tangents of the quartic, and whose inflexional tangent is the double tangent to the original quartic.*

One caution is necessary. The line through the external focus of an elliptic limaçon cuts the curve in two real points P, Q and two imaginary ones P', Q', or in four imaginary points. Now an imaginary projection, which converts the circular points into two real cusps, may convert the two imaginary points P', Q' into two real points, whilst P and Q still remain real. Hence we must be careful to take the tangents at the pair of points corresponding to P and Q or P' and Q', and not to P and P'. Similar observations apply to the reciprocal theorem.

371. *Every quartic having three biflecnodes may be projected into a lemniscate of Bernoulli.*

We have shown in § 170 that when a quartic has three biflecnodes two of them must be real and one complex, or one must be real and the other two imaginary. Hence in the two respective cases, the two real or the two imaginary biflecnodes must be projected into the circular points at infinity, and the curve will become a lemniscate.

372. Properties of quartics having three biflecnodes may also be deduced from those of the four-cusped hypocycloid, or the evolute of an ellipse. We have shown in § 332 that the reciprocal polar of the hypocycloid is $a^2 (x^2 + y^2) = x^2 y^2$, which if the axes are

turned through an angle $\frac{1}{4}\pi$ becomes $4a^2(x^2 + y^2) = (x^2 - y^2)^2$. By § 188 this curve has a complex biflecnode at the origin and a pair of real ones at infinity which lie on the lines $y = \pm x$, and if in (3) of § 355 we put $l = \infty$, $m = 1$, the resulting formulae project the two real biflecnodes into the circular points, and the curve projects into the lemniscate $4a^2(y^2 - x^2) = (x^2 + y^2)^2$. Applying this projection to the theorem at the end of § 339, the circle becomes a rectangular hyperbola, whence :—

If a pair of real tangents be drawn to a lemniscate from any point on the curve, the envelope of the chord of contact is a rectangular hyperbola whose centre is the real node of the lemniscate.

In the general case of any quartic with three biflecnodes, the locus is a conic; but when four *real* tangents can be drawn from a point on the quartic, care must be taken to select the two pairs which correspond to the two real or two imaginary tangents which can be drawn to the lemniscate.

373. Properties of quartics having two nodes and a cusp or three nodes may be deduced from those of bicircular quartics having the same singularities. In the first case the bicircular quartic must be the inverse of a parabola, and in the second case the inverse of a central conic.

374. Another class of simple forms may be deduced by the following method, which is one of general application. Let the triangle of reference be projected into an equilateral triangle in such a manner that the line $l\alpha + m\beta + n\gamma = 0$ becomes the line at infinity. Then if (α, β, γ), $(\alpha', \beta', \gamma')$ be corresponding points in the two planes, it follows that the line (l, m, n) is projected into the line $\alpha' + \beta' + \gamma' = 0$; hence we may take $\alpha' = l\alpha$, $\beta' = m\beta$, $\gamma' = n\gamma$, and the substitution of these values in the equation of the curve will furnish a simpler curve of the same species.

375. When a trinodal quartic is such that the tangents at each node together with the lines joining this node to the other two nodes form a harmonic pencil, the quartic can be projected into another curve in which the three nodes are situated at the vertices of an equilateral triangle and the lines bisecting the three nodal tangents intersect at the centre of gravity of the triangle. For this special class of quartics the theorems of §§ 192—4 may be at once proved by inspection; but although theorems connected

with this special class of quartics cannot be generalized for every trinodal quartic by projection, it is worth while to point out that the study of a special form may often suggest theorems which, although incapable of being proved in the general case by projection, are nevertheless true, and may be established by other methods. We may add that the lemniscate of Gerono, § 258, is one of the simplest quartics having a biflecnode and a tacnode; whilst the conchoid of Nicomedes, § 305, is one of the simplest forms of a quartic having a tacnode and one other double point, which may be a node or a cusp.

376. All the projective properties of binodal quartics, in which the two nodes are of the same kind, may be deduced from those of bicircular quartics; but if the nodes are different, as in the case of a quartic having an ordinary node and a flecnode, this method cannot be employed, but a special investigation is necessary. The projective properties of bicuspidal quartics may be deduced from those of cartesians.

377. With regard to quartics having a tacnode cusp or an oscnode; or a rhamphoid cusp or a tacnode with or without another double point, simple forms may be obtained by taking the singularity as the origin or projecting it to infinity. And when there are two singularities, both may be projected to infinity by performing this operation on the line joining them.

On a Special Quartic.

378. The general expression for a ternary quartic contains fifteen terms which may be arranged in five sets of three. The leading terms of each set are α^4; $\alpha^3\beta$; $\alpha^3\gamma$; $\alpha^2\beta^2$; $\alpha^2\beta\gamma$, constant multipliers being understood.

The first set equated to zero is the equation of a quartic having twelve points of undulation, four of which are real and the remaining eight are imaginary. The second set is the quartic whose properties will now be discussed, whilst the third set represents the same quartic differently situated with respect to the triangle of reference. The fourth set is the equation of a quartic having three biflecnodes. The fifth set represents four straight lines; whilst the sum of the fourth and fifth sets is the equation of a quartic having three double points.

379. The equation

$$l\alpha^3\beta + m\beta^3\gamma + n\gamma^3\alpha = 0 \dots\dots\dots\dots\dots(1)$$

has been considered by Klein in connection with a septimic transformation in the Theory of Functions*. Putting $\alpha = 0$, we obtain $\beta^3\gamma = 0$, which shows that C is a point of inflexion, and that BC is the tangent at C; hence A, B and C are points of inflexion, and CA, AB and BC are the tangents at these points. The peculiarity of this quartic is that there are three other real points of inflexion A', B', C' such that $A'B'$, $B'C'$, $C'A'$ are the tangents at these points, and that the equation of the quartic referred to $A'B'C'$ is of the form $\lambda\alpha'^3\gamma' + \mu\gamma'^3\beta' + \nu\beta'^3\alpha' = 0$; also a conic can be described through the six points A, B, C, A', B', C'. It will, however, be unnecessary to consider the general case, since the above theorems can be proved by investigating the special case of a quartic which is symmetrically situated with respect to an equilateral triangle, and then generalizing by projection.

380. The equation of the quartic may now be written

$$\alpha^3\beta + \beta^3\gamma + \gamma^3\alpha = 0 \dots\dots\dots\dots\dots(2),$$

and the equation of the circle circumscribing the triangle of reference is

$$\beta\gamma + \gamma\alpha + \alpha\beta = 0 \dots\dots\dots\dots\dots(3).$$

To find where (3) cuts (2), eliminate γ and put $\beta/\alpha = k$, and we shall obtain

$$\alpha\beta\left\{(1 + k)^3 - (1 + k)^2\, k^3 - k^2\right\} = 0 \dots\dots\dots\dots(4).$$

The first factor shows that the quartic passes through B and A; and the second factor gives the remaining five points of intersection of the circle and the quartic. This may be written in the form

$$(1 + k + k^2)(1 + 2k - k^2 - k^3) = 0 \dots\dots\dots\dots(5).$$

The factor $1 + k + k^2$, when equated to zero, gives the lines joining C to the circular points at infinity; whilst the equation

$$k^3 + k^2 = 2k + 1 \dots\dots\dots\dots\dots\dots(6)$$

* *Math. Annalen*, vol. XIV. p. 428.

gives the remaining three points A', B', C'; and we shall now show that these are points of inflexion.

Let the equation of CA' be $\alpha k = \beta$, where k is one of the roots of (6); then since A' lies on (3), we find that the equation of AA' is $\beta + (k+1)\gamma = 0$, whence the coordinates of A' are

$$\beta = k\alpha, \quad \gamma = -k\alpha/(1+k)\dots\dots\dots\dots(7).$$

Also the polar conic of any point $(\xi,\ \eta,\ \zeta)$ is

$$\alpha^2\xi\eta + \beta^2\eta\zeta + \gamma^2\zeta\xi + \alpha\beta\xi^2 + \beta\gamma\eta^2 + \gamma\alpha\zeta^2 = 0\dots\dots\dots(8),$$

and therefore the polar conic of A' is

$$k\alpha^2 - \frac{k^2\beta^2 + k\gamma^2}{1+k} + k^2\beta\gamma + \frac{k^2\gamma\alpha}{(1+k)^2} + \alpha\beta = 0\dots\dots\dots(9),$$

and the discriminant is

$$4\Delta = \frac{5k^4}{(1+k)^2} - k^5 + \frac{k^6}{(1+k)^5} + \frac{k}{1+k}\dots\dots\dots\dots(10).$$

Now (6) may be written

$$\frac{k}{(k+1)^2} = k - 1\dots\dots\dots\dots\dots\dots(11).$$

Using this in the third term of (10) we obtain

$$4\Delta = \frac{k}{k+1}\left(\frac{5k^3}{1+k} + 1 + k^3 - 3k^4\right).$$

Using (6) again, we obtain after some reduction

$$4\Delta = \frac{5k^2}{(k+1)^2}(k^3 + k^2 - 2k - 1)$$

$$= 0.$$

Whence the polar conic of A' breaks up into two straight lines, and therefore A' is a point of inflexion. In the same way it can be shown that B' and C' are points of inflexion; whence a circle can be described through all six points of inflexion.

From (7) it follows that the equation of a line through C' parallel to AA' is

$$\alpha(1+k) + k\beta = 0\dots\dots\dots\dots\dots(12),$$

and this line intersects the circumscribing circle at a point B'' such that $\gamma = k\beta$; whence the coordinates of B'' are

$$\alpha = - k\beta/(1 + k), \quad \gamma = k\beta \dots\dots\dots\dots\dots(13).$$

Now if in (6) we substitute $-(1 + k)/k$ for k, we shall find that it is satisfied; hence this quantity is another root of the cubic, which shows that B'' is one of the points of intersection of the circumscribing circle with the curve; hence B'' coincides with B'. In the same way it can be shown that the coordinates of C' are

$$\alpha = k\gamma, \quad \beta = - k\gamma/(1 + k) \dots\dots\dots\dots\dots(14),$$

and that CC' is parallel to $A'B$. Whence the three pairs of straight lines AA', CB'; BB', AC'; CC', BA' are respectively parallel to one another; also $-(1 + k)^{-1}$ is the third root of the cubic (6). Collecting our results, we find that the coordinates of A', B', C' are determined by the equations

$$\left. \begin{array}{ll} A', & \beta = k\alpha, \quad \gamma = - k\alpha/(1 + k) \\ B', & \alpha = - k\beta/(1 + k), \quad \gamma = k\beta \\ C', & \alpha = k\gamma, \quad \beta = - k\gamma/(1 + k) \end{array} \right\} \dots\dots\dots(15).$$

The equation of the tangent at A' is

$$\alpha \, (3\xi^2\eta + \zeta^3) + \beta \, (3\eta^2\zeta + \xi^3) + \gamma \, (3\zeta^2\xi + \eta^3) = 0.$$

Substituting the values of (ξ, η, ζ) from the first of (15), this becomes

$$\alpha \left\{ 3k - \frac{k^3}{(1 + k)^3} \right\} + \beta \left\{ 1 - \frac{3k^3}{1 + k} \right\} + \gamma \left\{ k^3 + \frac{3k^2}{(1 + k)^2} \right\} = 0 \dots\dots(16).$$

To prove that the tangent at A' passes through B', substitute for (α, β, γ) in (16) the coordinates of B' from the second of (15), and we obtain

$$\frac{k^4}{(1 + k)^4} - 3k^2 + 1 + k^4 + \frac{3k^3}{(1 + k)^2},$$

which by means of (6) and (11) may be shown to vanish. This shows that $A'B'$, $B'C'$, $C'A'$ are the tangents at the points of inflexion A', B', C'.

381. It remains to prove that the triangle $A'B'C'$ is equilateral.

Since BB' is parallel to AC', and $ABB'C'$ lie on a circle,

$$B'C'A = \pi - C'B'B = BAC';$$

whence $\qquad A + CAC' = C' + AC'A'.$

Also
$$A = C = BC'A$$
$$= BC'A' + AC'A'.$$

Whence
$$C' = BC'A' + CAC',$$

accordingly
$$BC'B' = CAC'$$
$$= CB'C' = CBC',$$

whence BC' is parallel to $B'C$ and AA'.   Accordingly
$$BC'A' = BAA' = ABC',$$

therefore
$$\tfrac{1}{3}\pi = B = CBC' + ABC'$$
$$= BC'B' + BC'A' = C',$$

whence the triangle $A'B'C'$ is equilateral.

382.   The quartic (1) is anautotomic; for its discriminant is equal to $26lmn$, which cannot vanish unless one of the constants l, m, n is zero, in which case the quartic splits up into a cubic and a straight line.

By tracing the symmetrical quartic, it can easily be seen that the points A, A', &c. are the only six real points of inflexion ; and also that the quartic has only three real double tangents which touch the curve at *real* points, and that the six real points of inflexion and the six points of contact of these double tangents lie on two concentric circles.   If in addition we eliminate γ between (2) and the line at infinity, we obtain $(\alpha^2 + \alpha\beta + \beta^2)^2 = 0$, which shows that this line is a double tangent whose points of contact are the circular points.   If therefore we generalize by a *real* projection, we obtain the theorem :—

The six real points of inflexion, and the six real points of contact of the three double tangents lie on two conics which touch one another at the two imaginary points where a fourth real double tangent touches the quartic.

The projection may be accomplished as follows.   Let the equilateral triangle ABC be projected into the triangle $A'B'C'$; and let (α, β, γ) and $(\alpha', \beta', \gamma')$ be the coordinates of two corresponding points P and P' in the two planes referred to the triangles ABC and $A'B'C'$ respectively; also let the projection be such that
$$\alpha = \alpha'/\lambda, \quad \beta = \beta'/\mu, \quad \gamma = \gamma'/\nu,$$

then the line at infinity and the circumscribing circle project in the line

$$\alpha'/\lambda + \beta'/\mu + \gamma'/\nu = 0,$$

and the conic
$$\lambda/\alpha' + \mu/\beta' + \nu/\gamma' = 0,$$

whilst the quartic projects into the curve

$$\alpha'^3\beta'/\lambda^3\mu + \beta'^3\gamma'/\mu^3\nu + \gamma'^3\alpha'/\nu^3\lambda = 0,$$

which will be identical with (1) if $\lambda = (m^3/l^9n)^{\frac{1}{28}}$ &c. &c.

ADDENDA AND CORRIGENDA.

I. In §§ 27—28 the number of tangents which can be drawn to a curve from a node or a cusp should be $m-4$ and $m-3$ respectively.

From any point O not on a curve the number of tangents is m. When O lies on the curve, two of the tangents coalesce with the tangent at O, leaving $m-2$. When O is a node, two pairs of tangents coalesce with the two nodal tangents, leaving $m-4$. When O is a cusp, three tangents coalesce with the cuspidal tangent, leaving $m-3$.

II. *The Cayleyan of a nodal cubic is a conic.*

The investigation of § 118 is not applicable to nodal cubics, since the canonical form has been used. We shall therefore prove that the Cayleyan of the logocyclic curve is a conic, whence by projection the theorem is true for any nodal cubic.

The equation of the curve is $x(x^2+y^2)=a(x^2-y^2)$; whence writing down the polar conic of any point (h, k) from (8) of § 130 it will be found that the equation of the Hessian is

$$5hk^2 + h^3 = a(k^2 - h^2)\dots\dots\dots\dots\dots\dots\dots(1);$$

also if ξ and η be the reciprocals of the intercepts which the polar conic cuts off from the axes

$$h = \frac{a}{3-2a\xi}, \qquad k = \frac{a\xi-2}{(3-2a\xi)\,\eta} \dots\dots\dots\dots (2).$$

Since the polar conic of every nodal cubic passes through the node, it follows that if (h, k) lie on the Hessian, the polar conic consists of the line $x\xi + y\eta = 1$ and a line through the origin. The envelope of the former line is the Cayleyan; whence eliminating (h, k) between (1) and (2), the tangential equation of the Cayleyan will be found to be

$$a^2 (\xi^2 - \eta^2) = a\xi + 2,$$

which represents a conic.

III. In connection with trinodal quartics, the following theorem due to Ferrers* may be noticed. His proof is instructive since it illustrates a method by which properties of trinodal quartics may be derived from conics. The theorem is :—

The six stationary tangents of a trinodal quartic touch a conic.

Let the equation of the trinodal quartic be

$$\lambda\beta^2\gamma^2 + \mu\gamma^2 a^2 + \nu a^2\beta^2 + 2a\beta\gamma (la + m\beta + n\gamma) = 0 \ldots \ldots (1),$$

and that of any tangent be

$$\xi a + \eta\beta + \zeta\gamma = 0 \ldots \ldots (2).$$

If one of the coordinates, say γ, be eliminated, the condition that (2) should be a stationary tangent is that three of the roots of the resulting equation in a/β should be equal. If therefore we write $1/a$, $1/\beta$, $1/\gamma$ for a, β, γ the conditions are the same as that the conics

$$\lambda a^2 + \mu\beta^2 + \nu\gamma^2 + 2l\beta\gamma + 2m\gamma a + 2na\beta = 0 \ldots \ldots (3)$$

and

$$\xi\beta\gamma + \eta\gamma a + \zeta a\beta = 0 \ldots \ldots (4)$$

should have a contact of the second order with one another. Writing these in the form $S = 0$, $S' = 0$, it follows that if k be determined so that the discriminant of $S + kS' = 0$ vanishes, the last equation will represent the three pairs of straight lines which can be drawn through the points of intersection of S and S'. By (4) of § 2, it follows that if Δ, Δ' be the discriminants of S and S' the discriminant of $S + kS'$ when equated to zero is

$$\Delta k^3 + 3\Theta k^2 + 3\Theta'k + \Delta' = 0 \ldots \ldots (5),$$

where

$$3\Theta = 2 (mn - l\lambda) \xi + 2 (nl - m\mu) \eta + 2 (lm - n\nu) \zeta,$$

$$3\Theta' = - \lambda\xi^2 - \mu\eta^2 - \nu\zeta^2 + 2l\eta\zeta + 2m\zeta\xi + 2n\xi\eta,$$

$$\Delta' = 2\xi\eta\zeta.$$

The conditions that the conics S and S' should have a contact of the second order with one another are that the three roots of (5) should be equal; which by (13) of § 7 are that

$$\frac{\Delta}{\Theta} = \frac{\Theta}{\Theta'} = \frac{\Theta'}{\Delta'} \ldots \ldots (6).$$

* *Quart. Journ.*, vol. XVIII. p. 73.

These equations are equivalent to $\Delta\Theta' = \Theta^2$ and $\Delta'\Theta = \Theta'^2$; the first of which, being an equation of the second degree in ξ, η, ζ, is the tangential equation of a conic which is touched by the six inflexional tangents. The second equation represents a curve of the fourth class, which also touches the six inflexional tangents.

IV. The duplication of the cube may be effected by means of the conchoid of Nicomedes in the following manner.

Let $AB = a$, $AC = b$ be two straight lines intersecting at right angles at A. On AC produced take a point D such that $AD = 2b$. On the side of AC remote from B take a point O, such that OAC is an isosceles triangle whose sides OA, OC are each equal to $\tfrac{1}{2}a$. Draw OM perpendicular to AC, and through A draw AE parallel to OD; draw OE perpendicular to AE, and with O as the node and AE as the asymptote describe the conchoid

$$r = OE \operatorname{cosec} \theta + \tfrac{1}{2}a$$

cutting CA in P. Join OP cutting AE in Q; then AP and OQ are the two required mean proportionals.

Since $AD = 2b$ and $PQ = \tfrac{1}{2}a$, and

$$OQ : PQ :: AD : AP,$$

we have

$$OQ \cdot AP = ab \dots\dots\dots\dots\dots\dots\dots(1).$$

Also

$$OM^2 = \tfrac{1}{4}(a^2 - b^2),$$
$$(OQ + \tfrac{1}{2}a)^2 = OP^2 = OM^2 + PM^2$$
$$= \tfrac{1}{4}(a^2 - b^2) + (AP + \tfrac{1}{2}b)^2,$$

whence

$$AP(AP + b) = OQ(OQ + a),$$

and therefore by (1)

$$AP^3 = a^2 b;$$

accordingly

$$\frac{AB}{AP} = \frac{AP}{OQ} = \frac{OQ}{AC}.$$

The three famous problems of antiquity were (i) the quadrature of the circle; (ii) the duplication of the cube, or the problem of finding two mean proportionals between two straight lines; and (iii) the trisection of an angle. The reader who desires to study the history of this subject is referred to the following works:—Leslie's *Geometrical Analysis*, edition 1821; Gow's *History of Greek Mathematics*; Lardner's *Algebraic Geometry*; Gregory's *Examples*; the Articles by De Morgan in the *Penny Cyclopædia*; and Klein's *Famous Problems of Elementary Geometry*.

THE END.

CAMBRIDGE: PRINTED BY J. AND C. F. CLAY, AT THE UNIVERSITY PRESS.